CONIC SECTIONS

TREATED GEOMETRICALLY

W. H. BESANT Sc. D. F.R.S.
FELLOW OF ST JOHN'S COLLEGE CAMBRIDGE

Published by

Hawk Press
4836/24, Ansari Road, Daryaganj
New Delhi - 110 002
Phones : 9643330713, 011-23278618, 011-35676207
E-mail: thehawkpress@gmail.com
www.thehawkpress.com

ISBN : 978-93-93971-52-4

PREFACE TO THE FIRST EDITION.

In the present Treatise the Conic Sections are defined with reference to a focus and directrix, and I have endeavoured to place before the student the most important properties of those curves, deduced, as closely as possible, from the definition.

The construction which is given in the first Chapter for the determination of points in a conic section possesses several advantages; in particular, it leads at once to the constancy of the ratio of the square on the ordinate to the rectangle under its distances from the vertices; and, again, in the case of the hyperbola, the directions of the asymptotes follow immediately from the construction. In several cases the methods employed are the same as those of Wallace, in the Treatise on Conic Sections, published in the *Encyclopaedia Metropolitana.*

The deduction of the properties of these curves from their definition as the sections of a cone, seems *à priori* to be the natural method of dealing with the subject, but experience appears to have shewn that the discussion of conics as defined by their plane properties is the most suitable method of commencing an elementary treatise, and accordingly I follow the fashion of the time in taking that order for the treatment of the subject. In Hamilton's book on *Conic Sections*, published in the middle of the last century, the properties of the cone are first considered, and the advantage of this method of commencing the subject, if the use of solid figures be not objected to, is especially shewn in the very general theorem of Art. (156). I have made much use of this treatise, and, in fact, it contains most of the theorems and problems which are now regarded as classical propositions in the theory of Conic Sections.

I have considered first, in Chapter I., a few simple properties of conics, and have then proceeded to the particular properties of each curve, commencing with the parabola as, in some respects, the simplest form of a conic section.

It is then shewn, in Chapter VI., that the sections of a cone by a plane produce the several curves in question, and lead at once to their definition as loci, and to several of their most important properties.

A chapter is devoted to the method of orthogonal projection, and another to the harmonic properties of curves, and to the relations of poles and polars, including the theory of reciprocal polars for the particular case in which the circle is employed as the auxiliary curve.

For the more general methods of projections, of reciprocation, and of anharmonic properties, the student will consult the treatises of Chasles, Poncelet, Salmon, Townsend, Ferrers, Whitworth, and others, who have recently developed, with so much fulness, the methods of modern Geometry.

I have to express my thanks to Mr R. B. Worthington, of St John's College, and of the Indian Civil Service, for valuable assistance in the constructions of Chapter XI., and also to Mr E. Hill, Fellow of St John's College, for his kindness in looking over the latter half of the proof-sheets.

I venture to hope that the methods adopted in this treatise will give a clear view of the properties of Conic Sections, and that the numerous Examples appended to the various Chapters will be useful as an exercise to the student for the further extension of his conceptions of these curves.

W. H. BESANT.

CAMBRIDGE,
March, 1869.

PREFACE TO THE NINTH EDITION.

In the preparation of this edition I have made many alterations and many additions. In particular, I have placed the articles on Reciprocal Polars in a separate chapter, with considerable expansions. I have also inserted a new chapter, on Conical Projections, dealing however only with real projections.

The first nine chapters, with the first set of miscellaneous problems, now constitute the elementary portions of the subject. The subsequent chapters may be regarded as belonging to higher regions of thought.

I venture to hope that this re-arrangement will make it easier for the beginner to master the elements of the subject, and to obtain clear views of the methods of geometry as applied to the conic sections.

A new edition, the fourth, of the book of solutions of the examples and problems has been prepared, and is being issued with this new edition of the treatise, with which it is in exact accordance.

W. H. BESANT.

December 14, 1894.

CONTENTS.

PAGE

INTRODUCTION 1

CHAPTER I.

THE CONSTRUCTION OF A CONIC SECTION, AND GENERAL PROPERTIES 3

CHAPTER II.

THE PARABOLA 20

CHAPTER III.

THE ELLIPSE 51

CHAPTER IV.

THE HYPERBOLA 88

CHAPTER V.

THE RECTANGULAR HYPERBOLA 125

CHAPTER VI.

THE CYLINDER AND THE CONE 135

CHAPTER VII.

The Similarity of Conics, the Areas of Conics, and the Curvatures of Conics 152

CHAPTER VIII.

Orthogonal Projections 165

CHAPTER IX.

Of Conics in General 174

CHAPTER X.

Ellipses as Roulettes and Glissettes 181

Miscellaneous Problems. I 189

CHAPTER XI.

Harmonic Properties, Poles and Polars 199

CHAPTER XII.

Reciprocal Polars 217

CHAPTER XIII.

The Construction of a Conic from Given Conditions . 231

CHAPTER XIV.

The Oblique Cylinder, the Oblique Cone, and the Conoids 245

CHAPTER XV.

Conical Projection 257

Miscellaneous Problems. II 269

CONIC SECTIONS.

INTRODUCTION.

DEFINITION.

If a straight line and a point be given in position in a plane, and if a point move in a plane in such a manner that its distance from the given point always bears the same ratio to its distance from the given line, the curve traced out by the moving point is called a Conic Section.

The fixed point is called the Focus, and the fixed line the Directrix of the conic section.

When the ratio is one of equality, the curve is called a Parabola.

When the ratio is one of less inequality, the curve is called an Ellipse.

When the ratio is one of greater inequality, the curve is called an Hyperbola.

These curves are called Conic Sections, because they can all be obtained from the intersections of a Cone by planes in different directions, a fact which will be proved hereafter.

It may be mentioned that a circle is a particular case of an ellipse, that two straight lines constitute a particular case of an hyperbola, and that a parabola may be looked upon as the limiting form of an ellipse or an hyperbola, under certain conditions of variation in the lines and magnitudes upon which those curves depend for their form.

The object of the following pages is to discuss the general forms and characters of these curves, and to determine their most important properties by help of the methods and relations developed in the first six books, and in the eleventh book of Euclid, and it will be found that, for this purpose, a knowledge of Euclid's Geometry is all that is necessary.

The series of demonstrations will shew the characters and properties which the curves possess in common, and also the special characteristics wherein they differ from each other; and the continuity with which the curves pass into each other will appear from the definition of a conic section as a Locus, or curve traced out by a moving point, as well as from the fact that they are deducible from the intersections of a cone by a succession of planes.

CHAPTER I.

PROPOSITION I.

The Construction of a Conic Section.

1. Take S as the focus, and from S draw SX at right angles to the directrix, and intersecting it in the point X.

DEFINITION. *This line SX, produced both ways, is called the Axis of the Conic Section.*

In SX take a point A such that the ratio of SA to AX is equal to the given ratio; then A is a point in the curve.

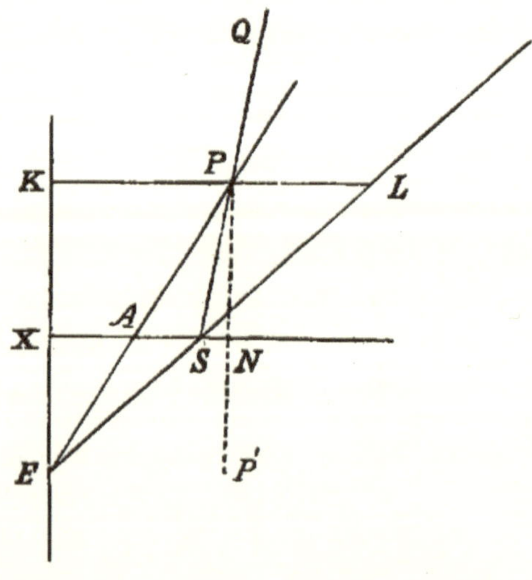

DEF. *The point A is called the Vertex of the curve.*

In the directrix EX take any point E, join EA, and ES, produce these lines, and through S draw the straight line SQ making with ES produced the same angle which ES produced makes with the axis SN.

Let P be the point of intersection of SQ and EA produced, and through P draw LPK parallel to NX, and intersecting ES produced in L, and the directrix in K.

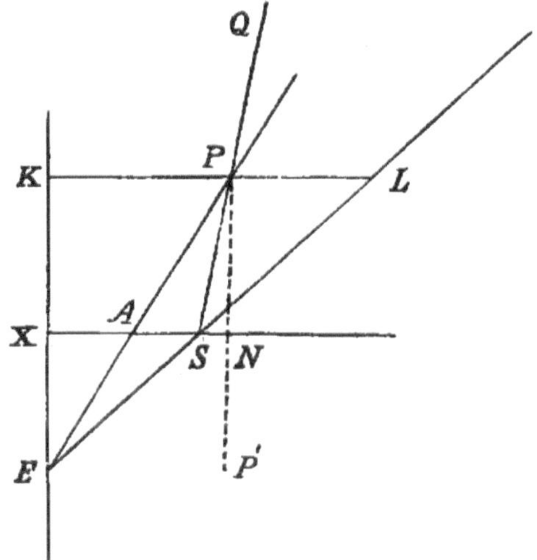

Then the angle PLS is equal to the angle LSN and therefore to PSL;

Hence $$SP = PL.$$

Also
$$PL : AS :: EP : EA$$
$$:: PK : AX;$$
$$\therefore PL : PK :: AS : AX;$$
and
$$\therefore SP : PK :: AS : AX.$$

The point P is therefore a point in the curve required, and by taking for E successive positions along the directrix we shall, by this construction, obtain a succession of points in the curve.

If E be taken on the upper side of the axis at the same distance from X, it is easy to see that a point P will be obtained below the axis, which will be similarly situated with regard to the focus and directrix. Hence it follows that the axis divides the curve into two similar and equal portions.

Another point of the curve, lying in the straight line KP, can be found in the following manner.

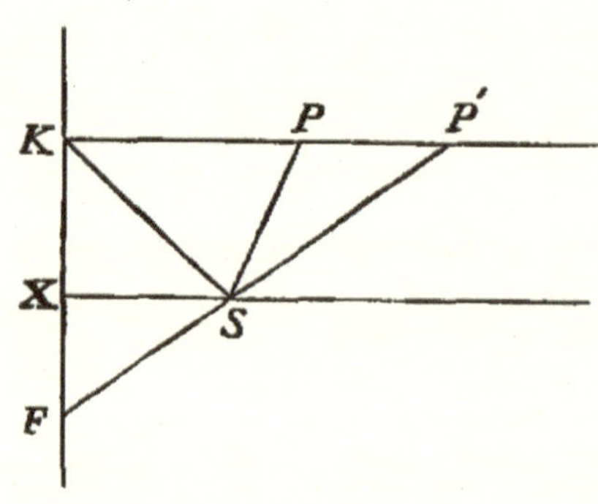

Through S draw the straight line FS making the angle FSK equal to KSP, and let FS produced meet KP produced in P'.

Then, since KS bisects the angle PSF,

$$SP' : SP :: P'K : PK;$$

$$\therefore SP' : P'K :: SP : PK,$$

and P' is a point in the curve.

2. DEF. *The Eccentricity. The constant ratio of the distance from the focus of any point in a conic section to its distance from the directrix is called the eccentricity of the conic section.*

The Latus Rectum. If E be so taken that EX is equal to SX, the angle PSN, which is double the angle LSN, and therefore double the angle ESX, is a right angle.

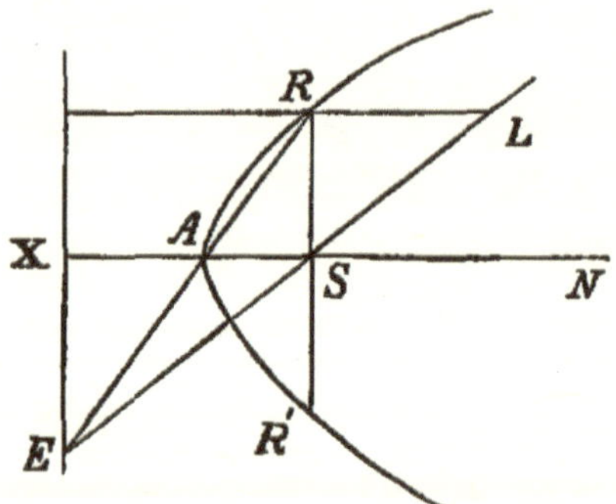

For, since $EX = SX$, the angle $ESX = SEX$, and, the angle SXE being a right angle, the sum of the two angles SEX, ESX, which is equal to twice ESX, is also equal to a right angle.

Calling R the position of P in this case, produce RS to R', so that $R'S = RS$; then R' is also a point in the curve.

DEF. *The straight line RSR' drawn through the focus at right angles to the axis, and intersecting the curve in R, and R', is called the Latus Rectum.*

It is hence evident that the form of a conic section is determined by its eccentricity, and that its magnitude is determined by the magnitude of the latus rectum, which is given by the relation

$$SR : SX :: SA : AX.$$

3. DEF. *The straight line PN (Fig. Art. 1), drawn from any point P of the curve at right angles to the axis, and intersecting the axis in N, is called the Ordinate of the point P.*

If the line PN be produced to P' so that $NP' = NP$, the line PNP' is a *double ordinate* of the curve.

The latus rectum is therefore the double ordinate passing through the focus.

DEF. *The distance AN of the foot of the ordinate from the vertex is called the Abscissa of the point P.*

DEF. *The distance SP is called the focal distance of the point P.*

It is also described as the radius vector drawn from the focus.

4. We have now given a general method of constructing a conic section, and we have explained the nomenclature which is usually employed. We proceed to demonstrate a few of the properties which are common to all the conic sections.

For the future the word conic will be employed as an abbreviation for conic section.

PROP. II. *If the straight line joining two points P, P′ of a conic meet the directrix in F, the straight line FS will bisect the angle between PS and P′S produced.*

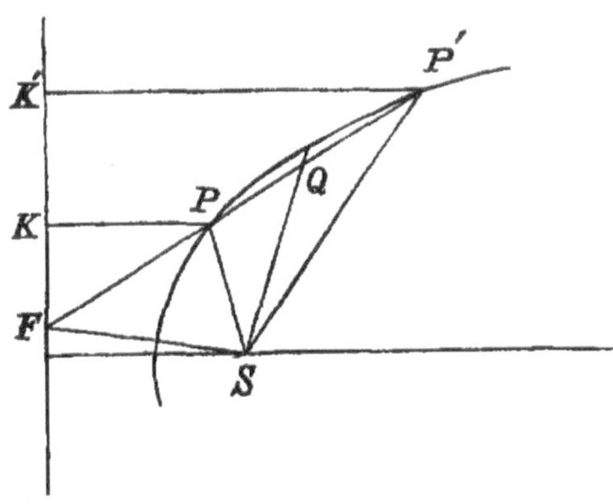

Draw the perpendiculars PK, $P'K'$ on the directrix.

Then
$$SP : SP' :: PK : P'K'$$
$$:: PF : P'F.$$

Therefore FS bisects the outer angle, at S, of the triangle PSP'. (Euclid VI., A.)

Cor. *If SQ bisect the angle PSP', it follows that FSQ is a right angle.*

5. Prop. III. *No straight line can meet a conic in more than two points.*

Employing the figure of Art. 4, let P be a point of the curve, and draw any straight line FP.

Join SF, draw SQ at right angles to SF, and SP' making the angle QSP' equal to QSP; then P' is a point of the curve.

For, since SF bisects the outer angle at S,

$$SP' : SP :: P'F : PF,$$

$$:: P'K' : PK$$

or

$$SP' : P'K' :: SP : PK,$$

and therefore, P' is a point of the curve, also, there is no other point of the curve in the straight line FPP'.

For suppose if possible P'' to be another point; then, as in Article (4), SQ bisects the angle PSP''; but SQ bisects the angle PSP'; therefore P'' and P' are coincident.

6. Prop. IV. *If QSQ' be a focal chord of a conic, and P any point of the conic, and if QP, $Q'P$ meet the directrix in E and F, the angle ESF is a right angle.*

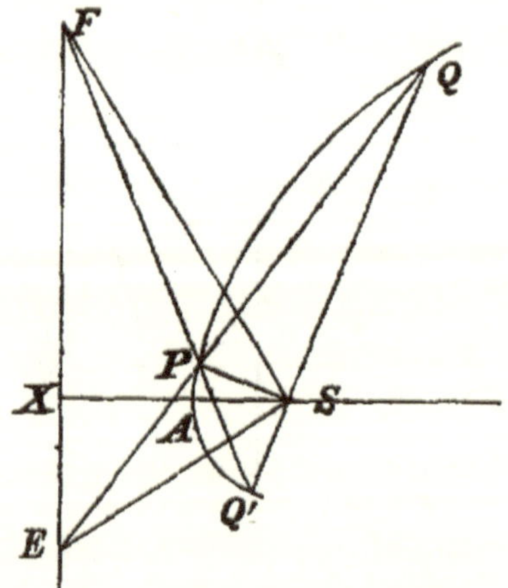

For, by Prop. II., SE bisects the angle PSQ', and SF bisects the angle PSQ;

hence it follows that ESF is a right angle.

This theorem will be subsequently utilised in the case in which the focal chord $Q'SQ$ is coincident with the axis of the conic.

7. Prop. V. *The straight lines joining the extremities of two focal chords intersect in the directrix.*

If PSp, $P'Sp'$ be the two chords, the point in which PP' meets the directrix is obtained by bisecting the angle PSP' and drawing SF at right angles to the bisecting line SQ. But this line also bisects the angle pSp'; therefore pp' also passes through F.

The line SF bisects the angle PSp', and similarly, if QS produced, bisecting the angle pSp', meet the directrix in F', the two lines Pp', $P'p$ will meet in F'. It is obvious that the angle FSF' is a right angle.

8. Prop. VI. *The semi-latus rectum is the harmonic mean between the two segments of any focal chord of a conic.*

Let PSP' be a focal chord, and draw the ordinates PN, $P'N'$.

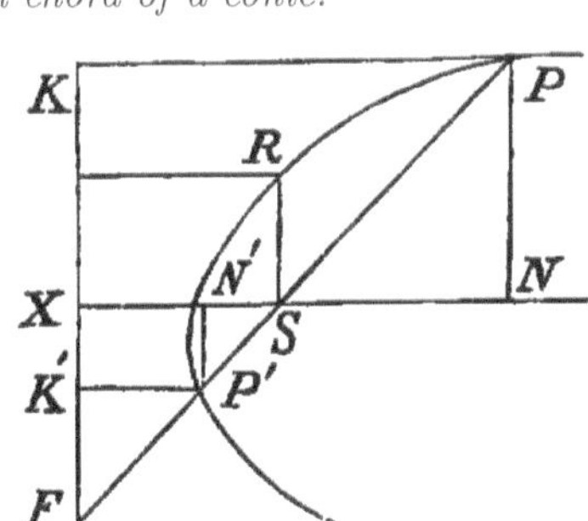

Then, the triangles SPN, $SP'N'$ being similar,

$$SP : SP' :: SN : SN'$$
$$:: NX - SX : SX - N'X$$
$$:: SP - SR : SR - SP',$$

since SP, SR, SP' are proportional to NX, SX, and $N'X$.

Cor. Since $SP : SP - SR :: SP \,.\, SP' : SP \,.\, SP' - SR \,.\, SP'$, and $SP' : SR - SP' :: SP \,.\, SP' : SR \,.\, SP - SP \,.\, SP'$, it follows that

$$SP \,.\, SP' - SR \,.\, SP' = SR \,.\, SP - SP \,.\, SP';$$

$$\therefore \quad SR \,.\, PP' = 2SP \,.\, SP'.$$

Hence, if PSP', QSQ' are two focal chords,

$$PP' : QQ' :: SP . SP' : SQ . SQ'.$$

9. Prop. VII. *A focal chord is divided harmonically at the focus and the point where it meets the directrix.*

Let PSP' produced meet the directrix in F, and draw PK, $P'K'$ perpendicular to the directrix, fig. Art. 8.

Then $PF : P'F :: PK : P'K'$

$$:: SP : SP'$$

$$:: PF - SF : SF - P'F;$$

that is, PF, SF, and $P'F$ are in harmonic progression, and the line PP' is divided harmonically at S and F.

10. *Definition of the Tangent to a curve.*

If a straight line, drawn through a point P of a curve, meet the curve again in P', and if the straight line be turned round the point P until the point P' approaches indefinitely near to P, the ultimate position of the straight line is the tangent to the curve at P.

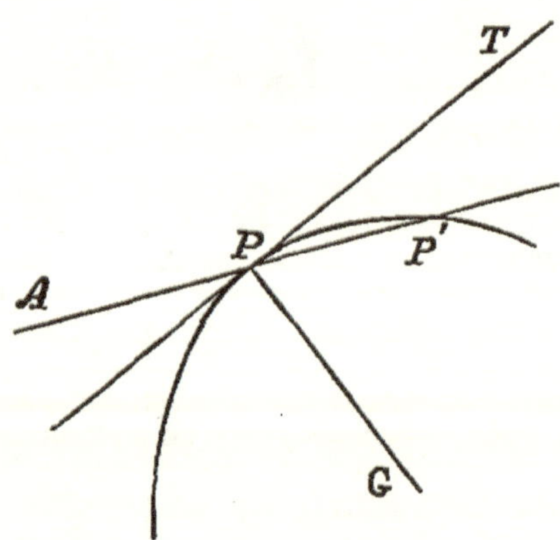

Thus, if the straight line APP' turn round P until the points P and P' coincide, the line in its ultimate position PT is the tangent at P.

Def. *The normal at any point of a curve is the straight line drawn through the point at right angles to the tangent at that point.*

Thus, in the figure, PG is the normal at P.

PROP. VIII. *The straight line, drawn from the focus to the point in which the tangent meets the directrix, is at right angles to the straight line drawn from the focus to the point of contact.*

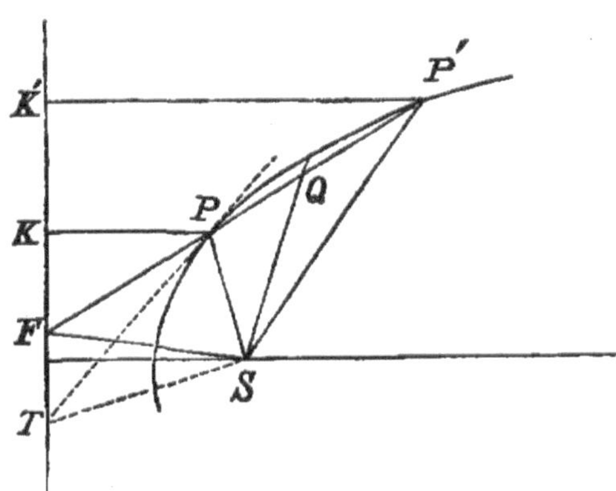

It is proved in Art. (4) that, if FPP' is a chord, and if SQ bisects the angle PSP', FSQ is a right angle.

Let the point P' move along the curve towards P; then, as P' approaches to coincidence with P, the straight line FPP' approximates to, and ultimately becomes, the tangent TP at P.

But when P' coincides with P, the line SQ coincides with SP, and the angle FSP, which is ultimately TSP, becomes a right angle.

Or, in other words, the portion of the tangent, intercepted between the point of contact and the directrix, subtends a right angle at the focus.

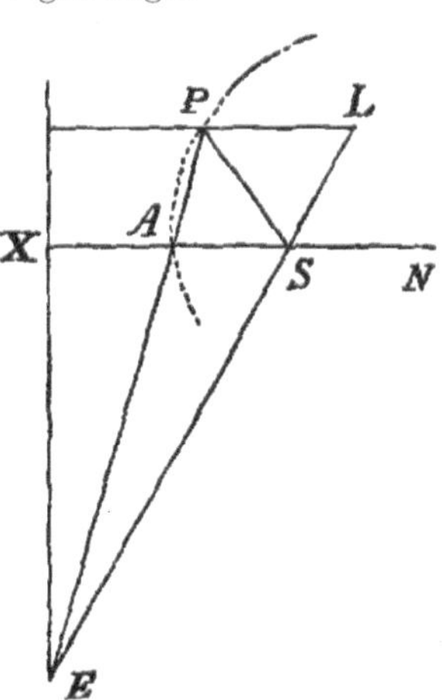

11. PROP. IX. *The tangent at the vertex is perpendicular to the axis.*

If a chord EAP be drawn through the vertex, and the point P be near the vertex, the angle PSA is small, and LSN, which is half the angle PSN, is nearly a right angle.

Hence it follows that when P approaches to coincidence with A, the point E moves off to an infinite distance and the line EAP, which is ultimately the tangent at A, becomes parallel to LSE, and is therefore perpendicular to AX.

12. PROP. X. *The tangents at the ends of a focal chord intersect on the directrix.*

For the line SF, perpendicular to SP, meets the directrix in the same point as the tangent at P; and, since SF is also at right angles to SP', the tangent at P' meets the directrix in the same point F.

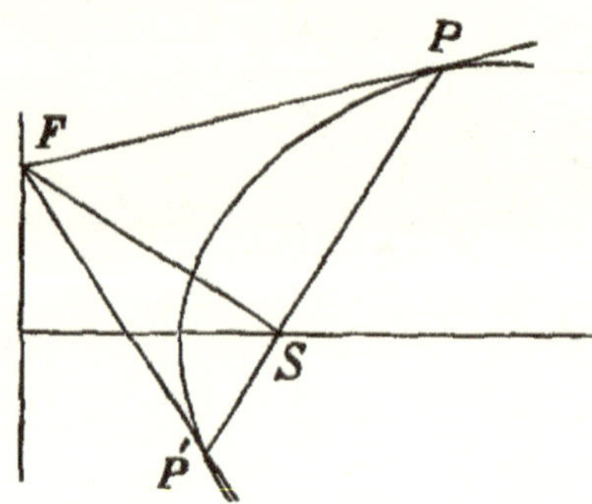

Conversely, if from any point F in the directrix tangents be drawn, the chord of contact, that is, the straight line joining the points of contact, will pass through the focus and will be at right angles to SF.

COR. Hence it follows that the tangents at the ends of the latus rectum pass through the foot of the directrix.

13. PROP. XI. *If a chord $P'P$ meet the directrix in F, and if the line bisecting the PSP' meet the curve in q and q', Fq and Fq' will be the tangents at q and q'.*

Taking the figure of Art. 7, the line SQ meets the curve in q and q', and, since SF is at right angles to SQ, it follows, from Art. 12, that Fq and Fq' are tangents.

Hence if from a point F in the directrix tangents be drawn, and also any straight line FPP' cutting the curve in P and P', the chord of contact will bisect the angle PSP'.

14. PROP. XII. *If the tangent at any point P of a conic intersect the directrix in F, and the latus rectum produced in D,*

$$SD : SF :: SA : AX.$$

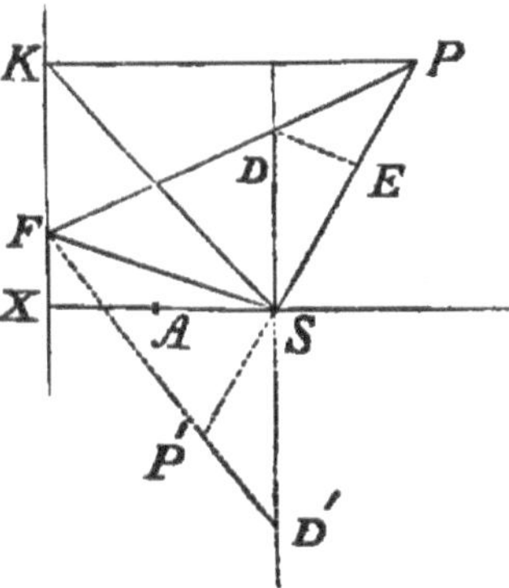

Join SK; then, observing that FSP and FKP are right angles, a circle can be described about $FSPK$, and therefore the angles SFD, SKP are equal.

Also the angle FSD

$$= \text{complement of } DSP$$
$$= SPK;$$

$\therefore$ the triangles FSD, SPK are similar, and

$$SD : SF :: SP : PK$$
$$:: SA : AX.$$

Cor. (1). If the tangent at the other end P' of the focal chord meet the directrix in D',

$$SD' : SF :: SA : AX;$$
$$\therefore SD = SD'.$$

Cor. (2). If DE be the perpendicular from D upon SP, the triangles SDE, SFX are similar, and

$$SE : SX :: SD : SF$$
$$:: SA : AX$$
$$:: SB : SX;$$

$\therefore SE$ is equal to SR, the semi-latus rectum.

15. Prop. XIII. *The tangents drawn from any point to a conic subtend equal angles at the focus.*

Let the tangents FTP, $F'TP'$ at P and P' meet the directrix in F and F' and the latus rectum in D and D'.

Join ST and produce it to meet the directrix in K;

then $$KF : SD :: KT : ST$$
$$:: KF' : SD'.$$

Hence $$KF : KF' :: SD : SD'$$
$$:: SF : SF' \text{ by Prop. XII.}$$

$\therefore$ the angles TSF, TSF' are equal.

But the angles FSP', $F'SP$ are equal, for each is the complement of FSF';

$\therefore$ the angles TSP, TSP' are equal.

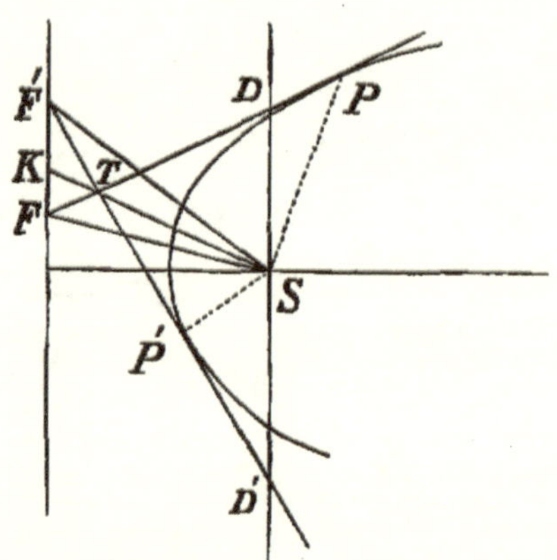

COR. Hence it follows that if perpendiculars TM, TM' be let fall upon SP and SP', they are equal in length.

For the two triangles TSM, TSM' have the angles TMS, TSM respectively equal to the angles $TM'S$, TSM', and the side TS common; and therefore the other sides are equal, and

$$TM = TM'.$$

16. PROP. XIV. *If from any point T in the tangent at a point P of a conic, TM be drawn, perpendicular to the focal distance SP, and TN perpendicular to the directrix,*

$$SM : TN :: SA : AX.$$

For, if PK be perpendicular to the directrix and SF be joined,

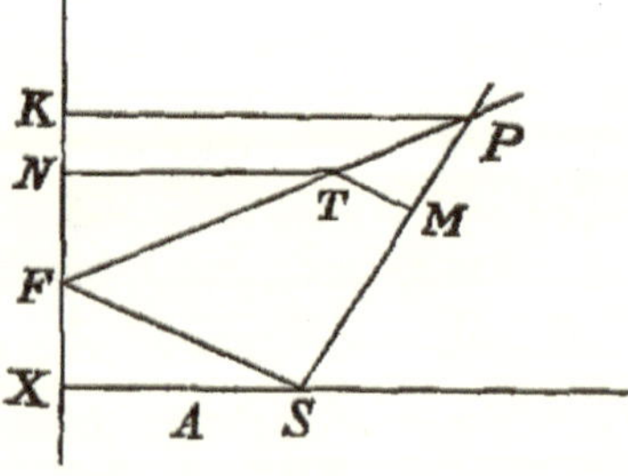

$$SM : SP :: TF : FP$$
$$:: TN : PK;$$
$$\therefore SM : TN :: SP : PK$$
$$:: SA : AX.$$

This theorem, which is due to Professor Adams, may be employed to prove Prop. XIII.

For if, in the figure of Art. (15), TM, TM' be the perpendiculars from T on SP and SP', and if TN be the perpendicular on the directrix, SM and SM' have each the same ratio to TN, and are therefore equal to one another.

Hence the triangles TSM, TSM' are equal in all respects, and the angle PSP' is bisected by ST.

17. Prop. XV. *To draw tangents from any point to a conic.*

Let T be the point, and let a circle be described about S as centre, the radius of which bears to TN the ratio of $SA : AX$; then, if tangents TM, TM' be drawn to the circle, the straight lines SM, SM', produced if necessary, will intersect the conic in the points of contact of the tangents from T.

18. Prop. XVI. *If PG, the normal at P, meet the axis of the conic in G,*

$$SG : SP :: SA : AX.$$

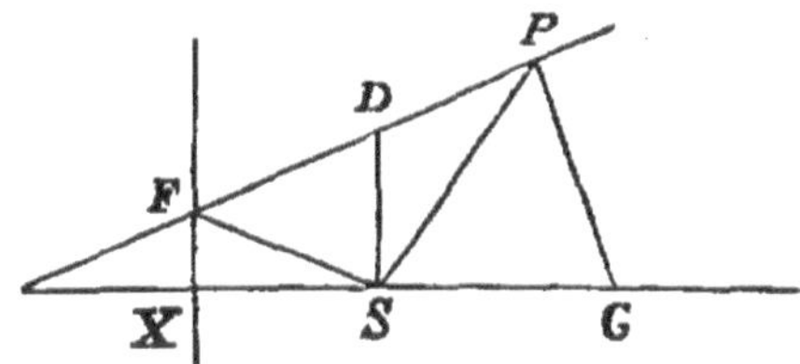

Let the tangent at P meet the directrix in F, and the latus rectum produced in D.

Then the angle SPG = the complement of $SPF = PFS$, and PSG = the complement of $FSX = FSD$;

$\therefore$ the triangles SFD, SPG are similar, and

$$SG : SP :: SD : SF :: SA : AX, \text{ by Prop. XII.}$$

19. Prop. XVII. *If from G, the point in which the normal at P meets the axis, GL be drawn perpendicular to SP, the length PL is equal to the semi-latus rectum.*

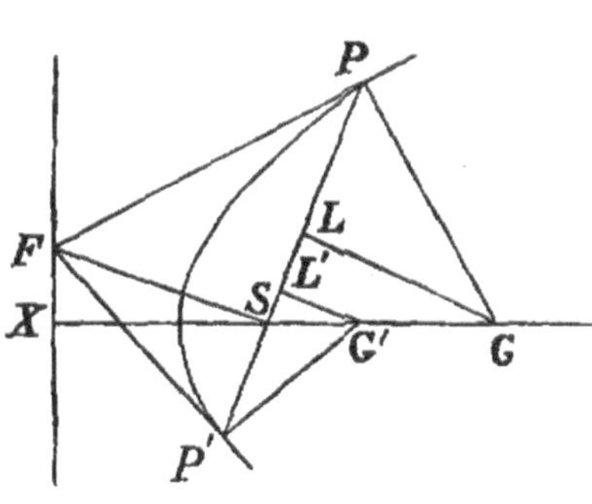

Let the tangent at P meet the directrix in F, and join SF.

Then PLG, PSF are similar triangles;

$$\therefore PL : LG :: SF : SP.$$

Also SLG and SFX are similar triangles;

$$\therefore LG : SX :: SG : SF.$$

Hence $$PL : SX :: SG : SP$$
$$:: SA : AX, \text{ Art. (18)},$$
but $$SR : SX :: SA : AX, \text{ Art. (2)};$$
$$\therefore PL = SR.$$

20. PROP. XVIII. *If from any point F in the directrix tangents be drawn, and also any straight line FPP' cutting the curve in P and P', the chord PP' is divided harmonically at F and its point of intersection with the chord of contact.*

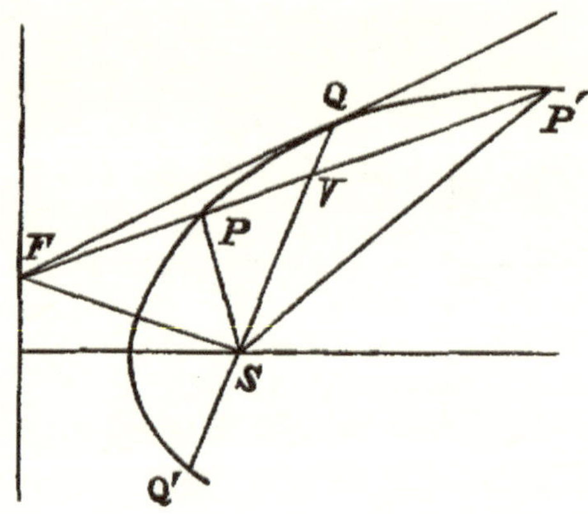

For, if QSQ' be the chord of contact, it bisects the angle PSP', (Prop. XI.), and $\therefore$, if V be the point of intersection of SQ and PP',

$$FP' : FP :: SP' : SP$$
$$:: P'V : PV$$
$$:: FP' - FV : FV - FP.$$

Hence FV is the harmonic mean between FP and FP'.

The theorems of this article and of Art. 9 are particular cases of more general theorems, which will appear hereafter.

21. PROP. XIX. *If a tangent be drawn parallel to a chord of a conic, the portion of this tangent which is intercepted by the tangents at the ends of the chord is bisected at the point of contact.*

Let PP' be the chord, TP, TP' the tangents, and EQE' the tangent parallel to PP'.

From the focus S draw SP, SP' and SQ, and draw TM, TM' perpendicular respectively to SP, SP'.

Also draw from E perpendiculars EN, EL, upon SP, SQ, and from E' perpendiculars $E'N'$, $E'L'$ upon SP' and SQ.

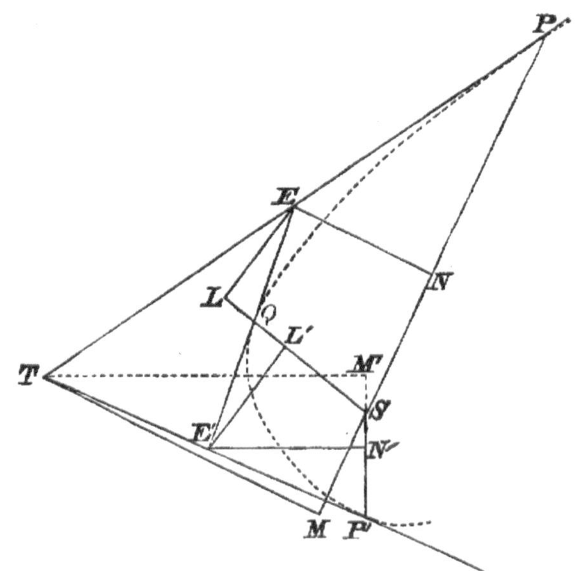

Then, since EE' is parallel to PP'

$$TP : EP :: TP' : E'P',$$

but $$TP : EP :: TM : EN,$$

and $$TP' : E'P' :: TM' : E'N';$$

$$\therefore TM : EN :: TM' : E'N';$$

but $TM = TM'$, Cor. Prop. XIII.;

$$\therefore EN = E'N'.$$

Again, by the same corollary,

$$EN = EL \text{ and } E'N' = E'L';$$

$$\therefore EL = E'L',$$

and, the triangles ELQ, $E'L'Q$ being similar,

$$EQ = E'Q.$$

COR. If TQ be produced to meet PP' in V,

$$PV : EQ :: TV : TQ,$$

and

$$P'V : E'Q :: TV : TQ;$$

$$\therefore PV = P'V,$$

that is, PP' is bisected in V.

Hence, if tangents be drawn at the ends of any chord of a conic, the point of intersection of these tangents, the middle point of the chord, and the point of contact of the tangent parallel to the chord, all lie in one straight line.

EXAMPLES.

1. Describe the relative positions of the focus and directrix, first, when the conic is a circle, and secondly, when it consists of two straight lines.

2. Having given two points of a conic, the directrix, and the eccentricity, determine the conic.

3. Having given a focus, the corresponding directrix, and a tangent, construct the conic.

4. If a circle passes through a fixed point and cuts a given straight line at a constant angle the locus of its centre is a conic.

5. If PG, pg, the normals at the ends of a focal chord, intersect in O, the straight line through O parallel to Pp bisects Gg.

6. Find the locus of the foci of all the conics of given eccentricity which pass through a fixed point P, and have the normal PG given in magnitude and position.

7. Having given a point P of a conic, the tangent at P, and the directrix, find the locus of the focus.

8. If PSQ be a focal chord, and X the foot of the directrix, XP and XQ are equally inclined to the axis.

9. If PK be the perpendicular from a point P of a conic on the directrix, and SK meet the tangent at the vertex in E, the angles SPE, KPE are equal.

10. If the tangent at P meet the directrix in F and the axis in T, the angles KSF, FTS are equal.

11. PSP' is a focal chord, PN, $P'N'$ are the ordinates, and PK, $P'K'$ perpendiculars on the directrix; if KN, $K'N'$ meet in L, the triangle LNN' is isosceles.

12. The focal distance of a point on a conic is equal to the length of the ordinate produced to meet the tangent at the end of the latus rectum.

13. The normal at any point bears to the semi-latus rectum the ratio of the focal distance of the point to the distance of the focus from the tangent.

14. The chord of a conic is given in length; prove that, if this length exceed the latus rectum, the distance from the directrix of the middle point of the chord is least when the chord passes through the focus.

15. The portion of any tangent to a conic, intercepted between two fixed tangents, subtends a constant angle at the focus.

16. Given two points of a conic, and the directrix, find the locus of the focus.

17. From any fixed point in the axis a line is drawn perpendicular to the tangent at P and meeting SP in R; the locus of R is a circle.

18. If the tangent at the end of the latus rectum meet the tangent at the vertex in T, $AT = AS$.

19. TP, TQ are the tangents at the points P, Q of a conic, and PQ meets the directrix in R; prove that RST is a right angle.

20. SR being the semi-latus rectum, if RA meet the directrix in E, and SE meet the tangent at the vertex in T,

$$AT = AS.$$

21. If from any point T, in the tangent at P, TM be drawn perpendicular to SP, and TN perpendicular to the transverse axis, meeting the curve in R, $SM = SR$.

22. If the chords PQ, $P'Q$ meet the directrix in F and F', the angle FSF' is half PSP'.

23. If PN be the ordinate, PG the normal, and GL the perpendicular from G upon SP,

$$GL : PN :: SA : AX.$$

24. If normals be drawn at the ends of a focal chord, a line through their intersection parallel to the axis will bisect the chord.

25. If a conic of given eccentricity is drawn touching the straight line FD joining two fixed points F and D, and if the directrix always passes through F, and the corresponding latus rectum always passes through D, find the locus of the focus.

26. If ST, making a constant angle with SP meet in T the tangent at P, prove that the locus of T is a conic having the same focus and directrix.

27. If E be the foot of the perpendicular let fall upon PSP' from the point of intersection of the normals at P and P',

$$PE = SP' \text{ and } P'E = SP.$$

28. If a circle be described on the latus rectum as diameter, and if the common tangent to the conic and circle touch the conic in P and the circle in Q, the angle PSQ is bisected by the latus rectum. (Refer to Cor. 2. Art. 14.)

29. Given two points, the focus, and the eccentricity, determine the position of the axis.

30. If a chord PQ subtend a constant angle at the focus, the locus of the intersection of the tangents at P and Q is a conic with the same focus and directrix.

31. The tangent at a point P of a conic intersects the tangent at the fixed point P' in Q, and from S a straight line is drawn perpendicular to SQ and meeting in R the tangent at P; prove that the locus of R is a straight line.

32. The circle is drawn with its centre at S, and touching the conic at the vertex A; if radii Sp, Sp' of the circle meet the conic in P, P', prove that PP', pp' intersect on the tangent at A.

33. Pp is any chord of a conic, PG, pg the normals, G, g being on the axis; GK, gk are perpendiculars on Pp; prove that $PK = pk$.

CHAPTER II.

THE PARABOLA.

DEF. *A parabola is the curve traced out by a point which moves in such a manner that its distance from a given point is always equal to its distance from a given straight line.*

Tracing the Curve.

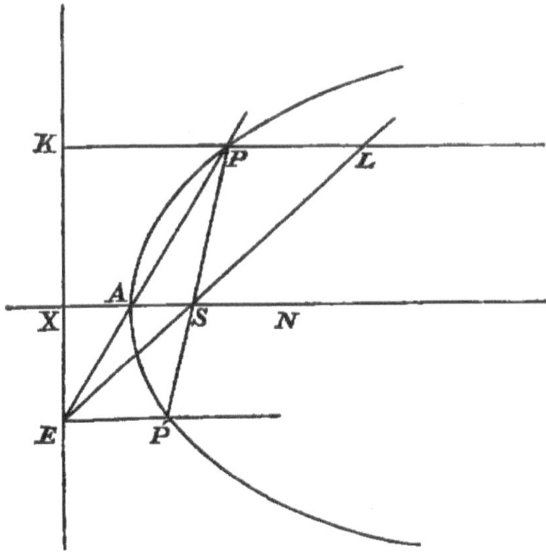

22. Let S be the focus, EX the directrix, and SX the perpendicular on EX. Then, bisecting SX in A, the point A is the vertex; and if, from any

point E in the directrix, EAP, ESL be drawn, and from S the straight line SP meeting EA produced in P, and making the angle PSL equal to LSN, we obtain, as in Art. (1), a point P in the curve.

For $$PL : PK :: SA : AX,$$

and $$\therefore PL = PK.$$

But $$SP = PL, \text{ and } \therefore SP = PK.$$

Again, drawing EP' parallel to the axis and meeting in P' the line PS produced, we obtain the other extremity of the focal chord PSP'.

For the angle $$ESP' = PSL = PLS$$
$$= SEP',$$

and $$\therefore SP' = P'E,$$

and P' is a point in the parabola.

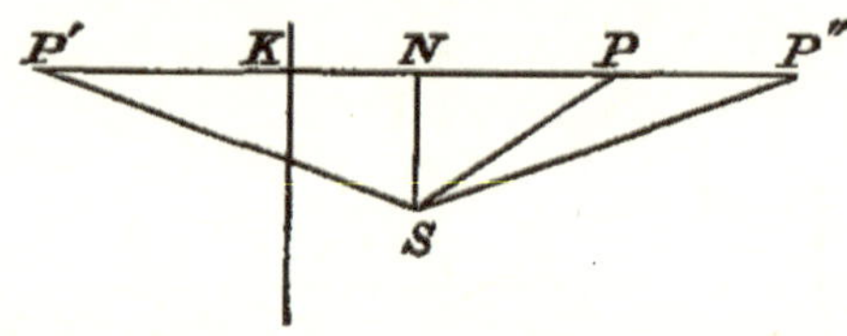

The curve lies wholly on the same side of the directrix; for, if P' be a point on the other side, and SN be perpendicular to $P'K$, SP' is greater than $P'N$, and therefore is greater than $P'K$.

Again, a straight line parallel to the axis meets the curve in one point only.

For, if possible, let P'' be another point of the curve in KP produced.

Then $$SP = PK \text{ and } SP'' = P''K$$
$$\therefore PP'' = SP'' - SP,$$

or $$PP'' + SP = SP'',$$

which is impossible.

23. PROP. I. *The distance from the focus of a point inside a parabola is less, and of a point outside is greater than its distance from the directrix.*

If Q be the point inside, let fall the perpendicular QPK on the directrix, meeting the curve in P.

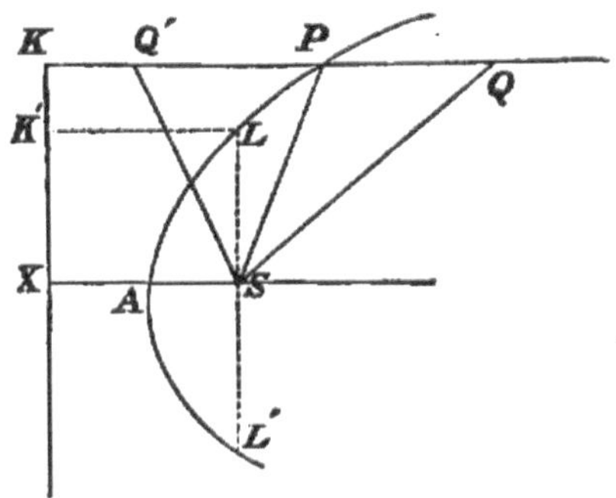

Then $SP + PQ > SQ$, but

$$SP + PQ = PK + PQ = QK,$$
$$\therefore SQ < QK.$$

If Q' be outside, and between P and K,

$$SQ' + PQ' > SP,$$
$$\therefore SQ' > Q'K.$$

If Q' lie in PK produced,

$$SQ' + SP > PQ',$$

and
$$\therefore SQ' > KQ'.$$

24. PROP. II. *The latus rectum* $= 4 \,.\, AS$.

For if, Fig. Art. 23, LSL' be the latus rectum, drawing LK' at right angles to the directrix, we have

$$LS = LK' = SX = 2AS,$$
$$\therefore LSL' = 4 \,.\, AS.$$

25. *Mechanical construction of the Parabola.*

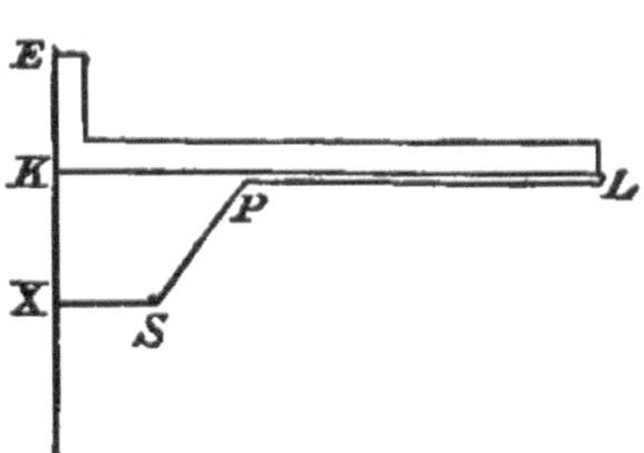

Take a rigid bar EKL, of which the portions EK, KL are at right angles to each other, and fasten a string to the end L, the length of which is LK. Then if the other end of the string be fastened to S, and the bar be made to slide along a fixed straight edge, EKX, a pencil at P, keeping the string stretched against the bar, will trace out a portion of a parabola, of which S is the focus, and EX the directrix.

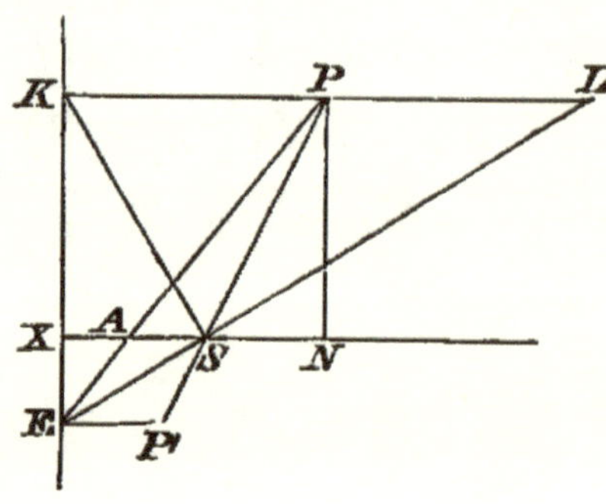

26. Prop. III. *If PK is the perpendicular upon the directrix from a point P of a parabola, and if PA meet the directrix in E, the angle KSE is a right angle.*

Join ES, and let KP and ES produced meet at L.

Since $SA = AX$, it follows that $PL = PK = SP$;

$\therefore$ P is the centre of the circle through K, S, and L, and the angle KSL is a right angle.

Therefore KSE is a right angle.

27. Prop. IV. *If PN is the ordinate of a point P of a parabola,*

$$PN^2 = 4AS \,.\, AN.$$

Taking the figure above,

$$PN : EX :: AN : AX$$

$$\therefore PN^2 : EX \,.\, KX :: 4AS \,.\, AN : 4AS^2.$$

But, since KSE is a right angle,

$$EX \,.\, KX = SX^2 = 4AS^2,$$

$$\therefore PN^2 = 4AS \,.\, AN.$$

Cor. If AN increases, and becomes infinitely large, PN increases and becomes infinitely large, and therefore the two portions of the curve, above and below the axis, proceed to infinity.

28. Prop. V. *If from the ends of a focal chord perpendiculars be let fall upon the directrix, the intercepted portion of the directrix subtends a right angle at the focus.*

For, if PA meet the directrix in E, and if the straight line through E perpendicular to the directrix meet PS in P', it is shewn, in Art. 22, that P' is the other extremity of the focal chord PS; and, as in Art. 26, KSE is a right angle.

29. Prop. VI. *The tangent at any point P bisects the angle between the focal distance SP and the perpendicular PK on the directrix.*

Let F be the point in which the tangent meets the directrix, and join SF.

We have shewn, (Art. 10) that FSP is a right angle, and, since $SP = PK$, and PF is common to the right-angled triangles SPF, KPF, it follows that these triangles are equal in all respects, and therefore the angle

$$SPF = FPK.$$

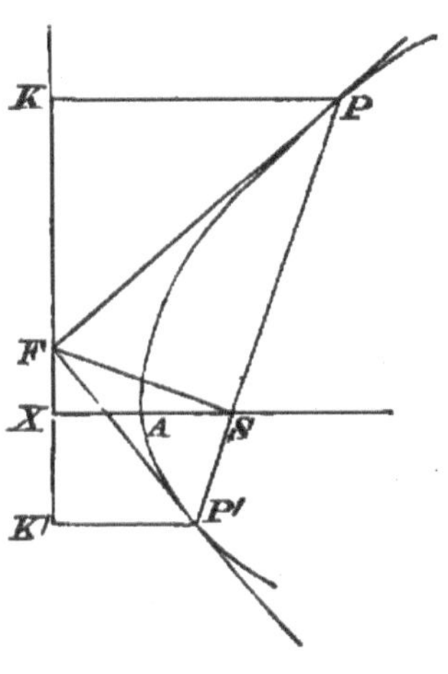

In other words, *the tangent at any point is equally inclined to the focal distance and the axis.*

Cor. It has been shewn, in Art. (12), that the tangents at the ends of a focal chord intersect in the directrix, and therefore, if PS produced meet the curve in P', FP' is the tangent at P', and bisects the angle between SP' and the perpendicular from P' on the directrix.

30. Prop. VII. *The tangents at the ends of a focal chord intersect at right angles in the directrix.*

Let PSP' be the chord, and PF, $P'F$ the tangents meeting the directrix in F.

Let fall the perpendiculars PK, $P'K'$, and join SK, SK'.

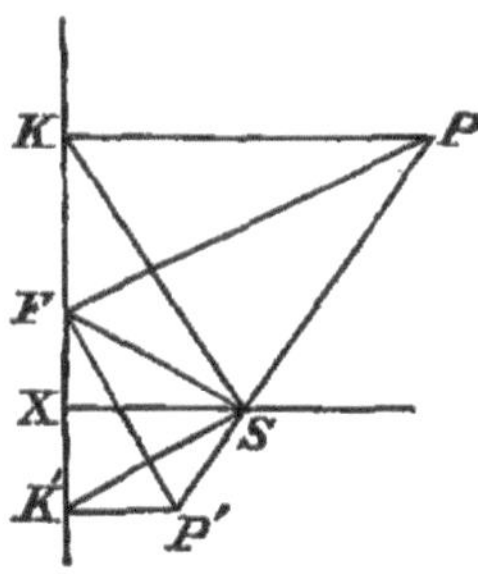

The angle $P'SK' = \frac{1}{2}P'SX$
$= \frac{1}{2}SPK = SPF$,

$\therefore SK'$ is parallel to PF,

and, similarly, SK is parallel to $P'F$.

But (Art. 28) KSK' is a right angle;

$\therefore PFP'$ is a right angle.

31. Prop. VIII. *If the tangent at any point P of a parabola meet the axis in T, and PN be the ordinate of P, then*

$$AT = AN.$$

Draw PK perpendicular to the directrix.

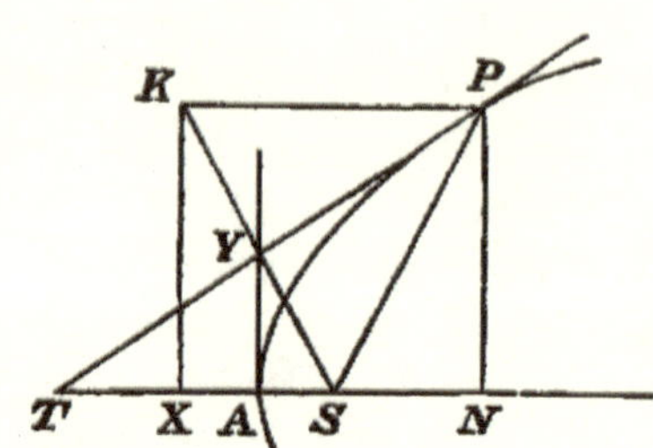

The angle $SPT = TPK$

$$= PTS,$$

$$\therefore ST = SP$$

$$= PK$$

$$= NX.$$

But $$ST = SA + AT,$$

and $$NX = AN + AX;$$

$$\therefore \text{ since } SA = AX,$$

$$AT = AN.$$

DEF. *The line NT is called the sub-tangent.*

The sub-tangent is therefore twice the abscissa of the point of contact.

32. PROP. IX. *The foot of the perpendicular from the focus on the tangent at any point P of a parabola lies on the tangent at the vertex, and the perpendicular is a mean proportional between SP and SA.*

Taking the figure of the previous article, join SK meeting PT in Y.

Then $SP = PK$, and PY is common to the two triangles SPY, KPY;

$$\text{also the angle } SPY = YPK;$$

$$\therefore \text{ the angle } SYP = PYK,$$

and SY is perpendicular to PT.

Also $SY = KY$, and $SA = AX$, $\therefore AY$ is parallel to KX.

Hence, AY is at right angles to AS, and is therefore the tangent at the vertex.

Again, the angle $SPY = STY = SYA$, and the triangles SPY, SYA are therefore similar;

$$\therefore SP : SY :: SY : SA,$$

$$\text{or } SY^2 = SP \,.\, SA.$$

33. PROP. X. *In the parabola the subnormal is constant and equal to the semi-latus rectum.*

DEF. *The distance between the foot of the ordinate of P and the point in which the normal at P meets the axis is called the subnormal.*

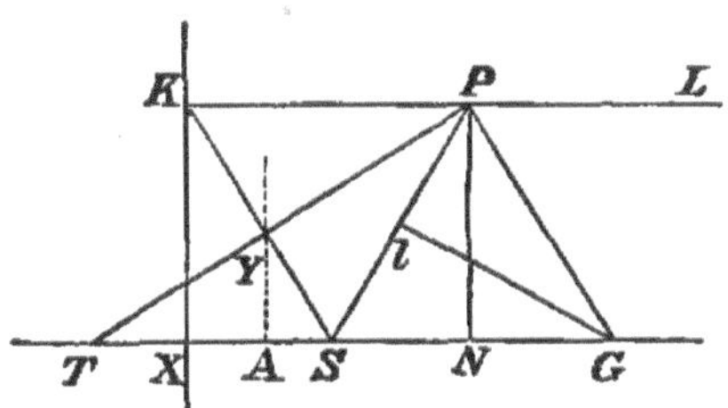

In the figure PG is the normal and PT the tangent.

It has been shewn that the angle SPK is bisected by PT, and hence it follows that SPL is bisected by PG,

and that the angle $\qquad SPG = GPL = PGS$;

hence
$$\begin{aligned} SG &= SP = ST \\ &= SA + AT = SA + AN \\ &= 2AS + SN; \end{aligned}$$

$\therefore$ the subnormal $NG = 2AS$.

34. Cor. If Gl be drawn perpendicular to SP,
the angle GPl = the complement of SPT,
= the complement of STP,
= PGN,

and the two right-angled triangles GPN, GPl have their angles equal and the side GP common; hence the triangles are equal, and

$$\begin{aligned} Pl &= NG = 2AS \\ &= \text{the semi-latus rectum.} \end{aligned}$$

It has been already shewn, (Art. 19), that this property is a general property of all conics.

35. Prop. XI. *To draw tangents to a parabola from an external point.*

For this purpose we may employ the general construction given in Art. (17), or, for the special case of the parabola, the following construction.

Let Q be the external point, join SQ, and upon SQ as diameter describe a circle intersecting the tangent at the vertex in Y and Y'. Join YQ, $Y'Q$; these are tangents to the parabola.

Draw SP, so as to make the angle YSP equal to YSA, and to meet YQ in P, and let fall the perpendicular PN upon the axis.

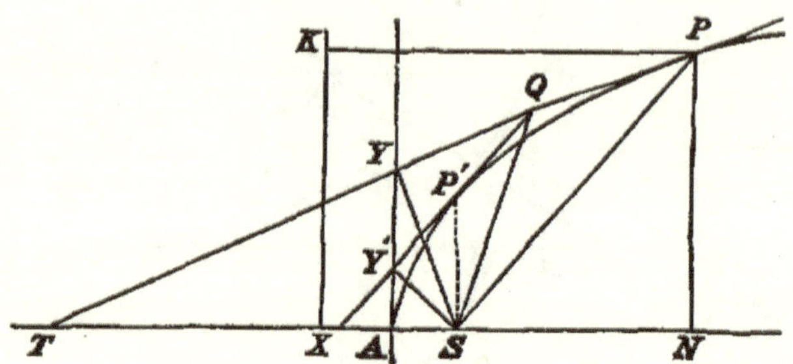

Then, SYQ is a right angle, since it is the angle in a semicircle, and, T being the point in which QY produced meets the axis, the two triangles SYP, SYT are equal in all respects;

$$\therefore SP = ST, \text{ and } YT = YP.$$

But AY is parallel to PN;

$$\therefore AT = AN.$$

Hence
$$\begin{aligned} SP &= ST = SA + AT \\ &= AX + AN \\ &= NX, \end{aligned}$$

and P is a point in the parabola.

Moreover, if PK be perpendicular to the directrix, the angle $SPY = STP = YPK$, and PY is the tangent at P. (Art. 29.)

Similarly, by making the angle $Y'SP'$ equal to ASY' we obtain the point of contact of the other tangent QY'.

36. PROP. XII. *If from a point Q tangents QP, QP' be drawn to a parabola, the two triangles SPQ, SQP', are similar, and SQ is a mean proportional between SP and SP'.*

Produce PQ to meet the axis in T, and draw SY, SY' perpendicularly on the tangents. Then Y and Y' are points in the tangent at A.

The angle
$$\begin{aligned} SPQ &= STY \\ &= SYA \\ &= SQP', \end{aligned}$$

since S, Y', Y, Q are points on a circle, and SYA, SQP' are in the same segment.

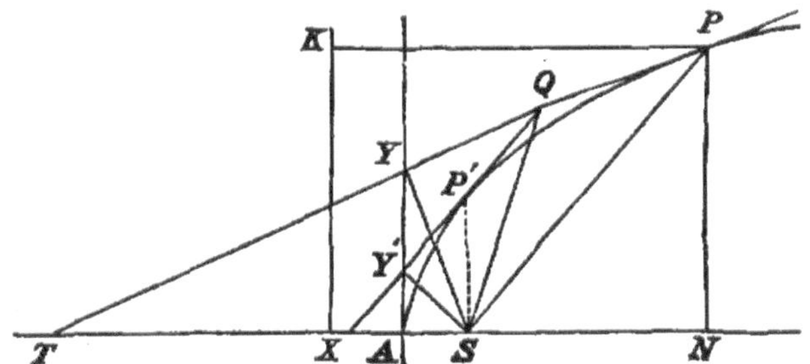

Also, by the theorem of Art. (15), the angle

$$PSQ = QSP';$$

therefore the triangles PSQ, QSP' are similar, and

$$SP : SQ :: SQ : SP'.$$

37. From the preceding theorem the following, which is often useful, immediately follows.

If from any points in a given tangent of a parabola, tangents be drawn to the curve, the angles which these tangents make with the focal distances of the points from which they are drawn are all equal.

For each of them by the theorem, is equal to the angle between the given tangent and the focal distance of the point of contact.

Hence it follows that the locus of the intersection of a tangent to a parabola with a straight line drawn through the focus meeting it at a constant angle is a straight line.

For if QP be the moveable tangent, the angle $SQP = SP'Q$, and therefore, if SQP is constant, $SP'Q$ is a given angle. The point P' is therefore fixed, and the locus of Q is the tangent $P'Q$.

38. Since the two triangles PSQ, QSP' are similar, we have

$$PQ : P'Q :: SP : SQ$$

and

$$PQ : P'Q :: SQ : SP',$$

$$\therefore PQ^2 : P'Q^2 :: SP : SP';$$

that is, the squares of the tangents from any point are proportional to the focal distances of the points of contact.

This will be found to be a particular case of a subsequent Theorem, given in Art. 51.

39. Prop. XIII. *The external angle between two tangents is half the angle subtended at the focus by the chord of contact.*

Let the tangents at P and P' intersect each other in Q and the axis ASN in T and T'.

Join SP, SP'; then the angles SPT, STP are equal, and $\therefore$ STP is half the angle PSN; similarly $ST'P'$ is half $P'SN$.

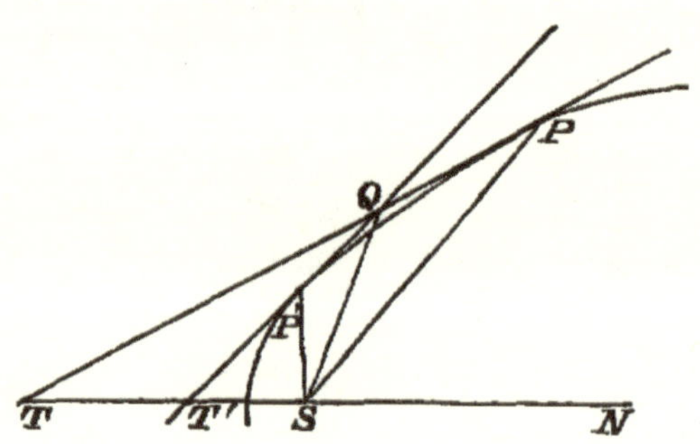

But TQT' is equal to the difference between STP and $ST'P'$, and is therefore equal to half the difference between PSN and $P'SN$, that is to half the angle PSP'.

Hence, joining SQ, TQT' is equal to each of the angles PSQ, $P'SQ$.

40. Prop. XIV. *The tangents drawn to a parabola from any point make the same angles, respectively, with the axis and the focal distance of the point.*

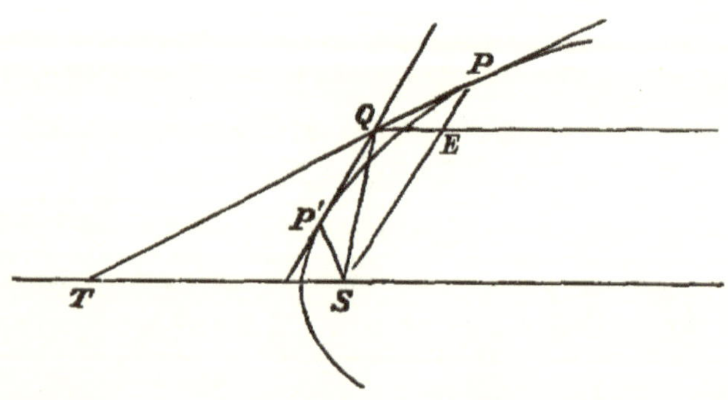

Let QP, QP' be the tangents; join SP, and draw QE parallel to the axis, and meeting SP in E.

Then, if PQ meet the axis in T, the angle

$$EQP = STP = SPQ$$
$$= SQP'. \quad \text{(Art. 37.)}$$

i.e. QP and QP' respectively make the same angles with the axis and with QS.

41. Conceive a parabola to be drawn passing through Q, having S for its focus, SN for its axis, and its vertex on the same side of S as the vertex A of the given parabola. Then the normal at Q to this new parabola bisects the angle SQE; therefore the angles which QP and QP' make with the normal at Q are equal.

Hence the theorem,

If from any point in a parabola, tangents be drawn to a confocal and co-axial parabola, the normal at the point will bisect the angle between the tangents.

If we produce SP to any point p, and take St equal to Sp, pt will be the tangent at p to the confocal and co-axial parabola passing through p.

Hence the theorem,

If parallel tangents be drawn to a series of confocal and co-axial parabolas, the points of contact will lie in a straight line passing through the focus.

In these enunciations the words co-axial and confocal are intended to imply, not merely the coincidence of the axes, but also that the vertices of the two parabolas are on the same side of their common focus.

The reason for this will appear when we shall have discussed the analogous property of the ellipse.

42. If two confocal parabolas have their axes in the same straight line, and their vertices on opposite sides of the focus, they intersect at right angles.

For the angle $TPS = \frac{1}{2}PST'$,

and $T'PS = \frac{1}{2}PST$,

$$\therefore TPT' = \tfrac{1}{2}(PST + PST') = \text{a right angle.}$$

It will be noticed that, in this case, the common chord PQ is equidistant from the directrices.

For the distance of P from each directrix is equal to SP.

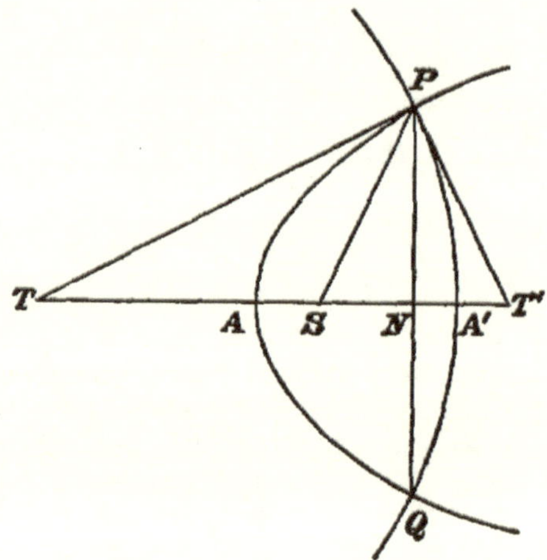

43. Prop. XV. *The circle passing through the points of intersection of three tangents passes also through the focus.*

Let Q, P, Q' be the three points of contact, and F, T, F' the intersections of the tangents.

In Art. (36) it has been shewn that, if FP, FQ be tangents, the angle

$$SQF = SFP.$$

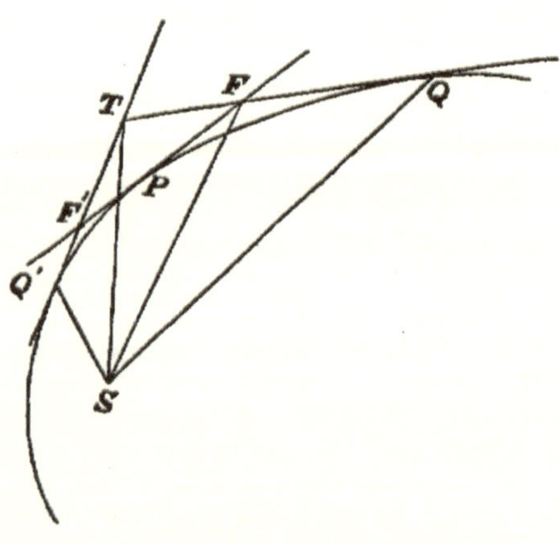

Similarly TQ, TQ' being tangents, the angle

$$SQT = STQ',$$

hence the angle $$SFF' \text{ or } SFP = SQT,$$
$$= STF',$$

and a circle can be drawn through S, F, T, and F'.

44. Def. *A straight line drawn parallel to the axis through any point of a parabola is called a diameter.*

Prop. XVI. *If from any point T tangents TQ, TQ' be drawn to a parabola, the point T is equidistant from the diameters passing through Q and Q', and the diameter drawn through the point T bisects the chord of contact.*

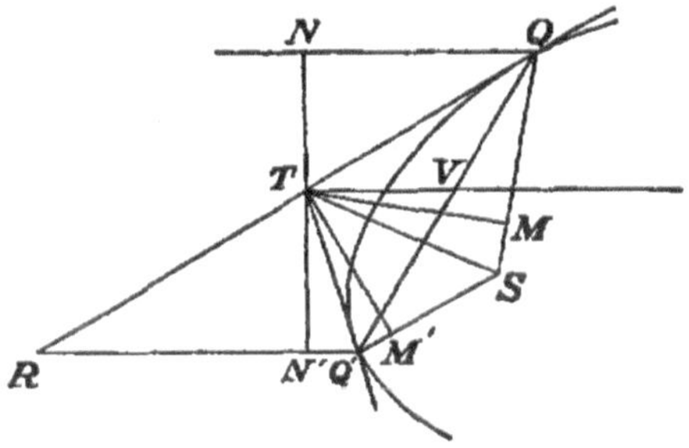

Join SQ, SQ', and draw TM, TM' perpendicular respectively to SQ and SQ'.

Also draw NTN' perpendicular to the diameters through Q and Q', and meeting those diameters in N and N'.

Then, since TS bisects the angle QSQ',

$$TM = TM';$$

and, since TQ bisects the angle SQN,

$$TN = TM.$$

Similarly $$TN' = TM',$$

$$\therefore TN = TN'.$$

Again, join QQ', and draw the diameter TV meeting QQ' in V; also let QT produced meet $Q'N'$ in R;

then $$QV : VQ' :: QT : TR$$
$$:: TN : TN',$$

since the triangles QTN, RTN' are similar;

$$\therefore QV = VQ'.$$

Hence *the diameter through the middle point of a chord passes, when produced, through the point of intersection of the tangents at the ends of the chord.*

It should be noticed that any straight line drawn through T and terminated by QN and $Q'N'$ is bisected at T.

45. PROP. XVII. *Any diameter bisects all chords parallel to the tangent at its extremity, and passes through the point of intersection of the tangents at the ends of any of these chords.*

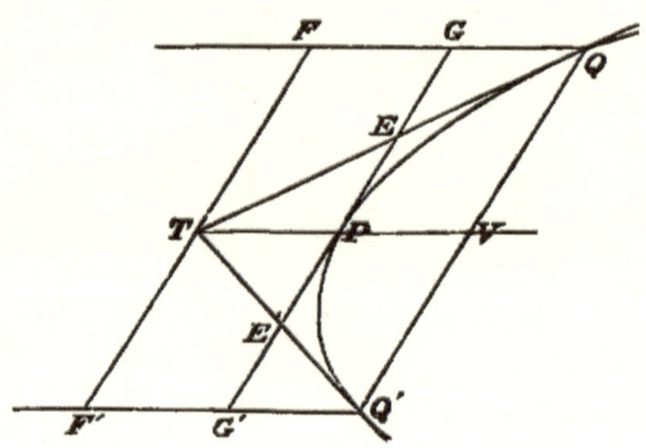

Let QQ' be a chord parallel to the tangent at P, and through the point of intersection T of the tangents at Q and Q' draw FTF' parallel to QQ' and terminated at F and F' by the diameters through Q and Q'.

Let the tangent at P meet TQ, TQ' in E and E', and QF, $Q'F'$ in G and G'.

Then
$$EG : TF :: EQ : TQ$$
$$:: E'Q' : TQ'$$
$$:: E'G' : TF'.$$

But $TF = TF'$, since (Art. 44) T is equidistant from QG and $Q'G'$,
$$\therefore EG = E'G'.$$

Also, $EP = EG$, since E is equidistant from QG and PV, the diameter at P.
$$\therefore EP = E'P \text{ and } GP = PG',$$
and
$$\therefore QV = VQ'.$$

Again, since T, P, V are each equidistant from the parallel straight lines QF, $Q'F'$, it follows that TPV is a straight line, or that the diameter VP passes through T.

We have shewn that GE, EP, PE', $E'G'$ are all equal, and we hence infer that

$$EE' = \tfrac{1}{2}GG' = \tfrac{1}{2}QQ',$$

and consequently that $TP = \frac{1}{2}TV$, or that $TP = PV$.

Hence *it appears, that the diameter through the point of intersection of a pair of tangents passes through the point of contact of the tangent parallel to the chord of contact, and also through the middle point of the chord of contact; and that the portion of the diameter between the point of intersection of the tangents and the middle point of the chord of contact is bisected at the point of contact of the parallel tangent.*

We may observe that in proving that EE' is bisected at P, we have demonstrated a theorem already shewn (Art. 21) to be true for all conics.

46. When the point T is on the directrix, QTQ' is a right angle.

If then Qq is the chord which is normal at Q, it is parallel to the tangent TQ', and is therefore bisected by the diameter $Q'U$ through Q'.

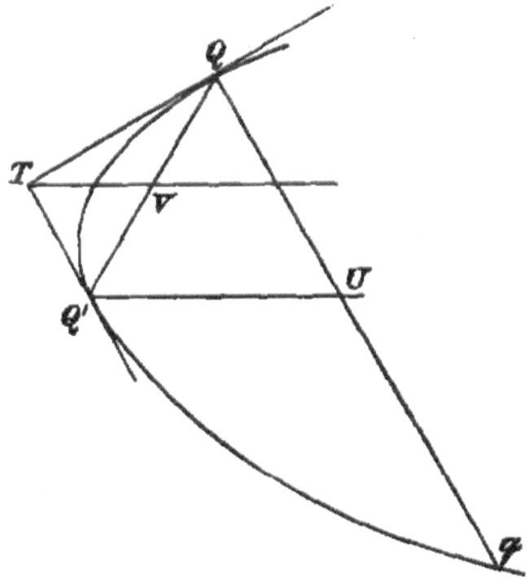

Since QU is bisected by TV, it follows that

$$Qq = 4TQ',$$

i.e. *the length of a normal chord is four times the portion of the parallel tangent between the directrix and the point of contact.*

47. Def. *The line QV, parallel to the tangent at P, and terminated by the diameter PV, is called an ordinate of that diameter, and QQ' is the double ordinate. The point P, the end of the diameter, is called the vertex of the diameter, and the distance PV is called the abscissa of the point Q.*

We have seen that tangents at the ends of any chord intersect in the diameter which bisects the chord, and that the distance of this point from the vertex is equal to the distance of the vertex from the middle point of the chord.

Def. *The chord through the focus parallel to the tangent at any point is called the parameter of the diameter passing through the point.*

Prop. XVIII. *The parameter of any diameter is four times the focal distance of the vertex of that diameter.*

Let P be the vertex, and QSQ' the parameter, T the point of intersection of the tangents at Q and Q', and FPF' the tangent at P.

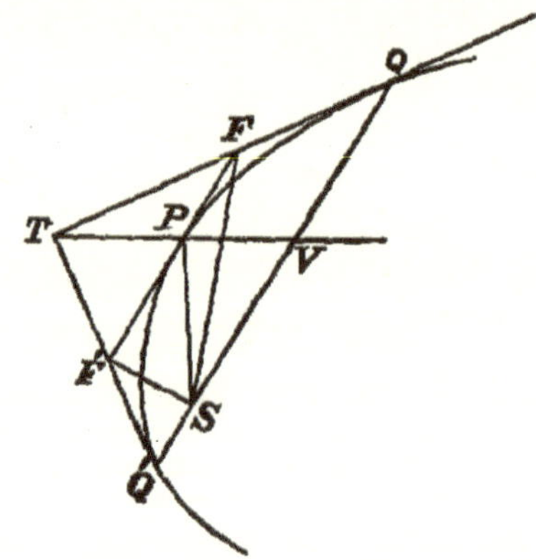

Then, since FS and $F'S$ bisect respectively the angles PSQ, PSQ', FSF' is a right angle, and, P being the middle point of FF', $SP = PF = PF'$.

Hence QQ', which is double FF', is four times SP.

48. Prop. XIX. *If QVQ' be a double ordinate of a diameter PV, QV is a mean proportional between PV and the parameter of P.*

Let FPF' be the tangent at P, and draw the parameter through S meeting PV in U.

The angle $SUT = FPU = SPF'$ (Art. 29), and, since the angles SFQ, SPF are equal (Art. 36), it follows that the angles SFT, SPF' are equal;

$\therefore SUT = SFT$, and U is a point in the circle passing through $SFTF'$.

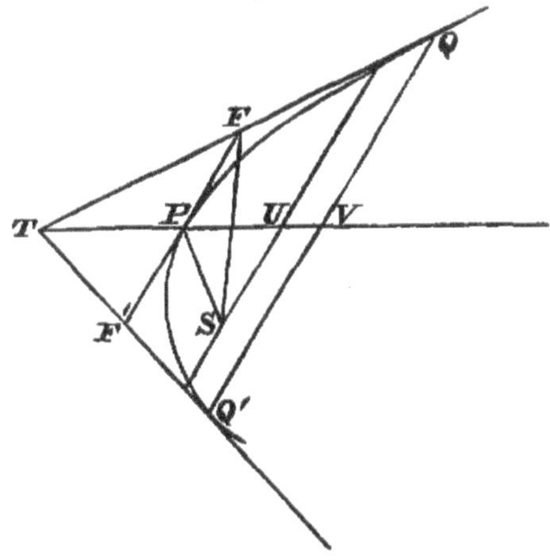

Hence, QV being twice PF,

$$QV^2 = 4PF^2 = 4PU \,.\, PT;$$

but $$PU = SP,$$

for the angle $$SUP = FPU = SPF' = PSU;$$

and $$PT = PV,$$

$$\therefore QV^2 = 4SP \,.\, PV.$$

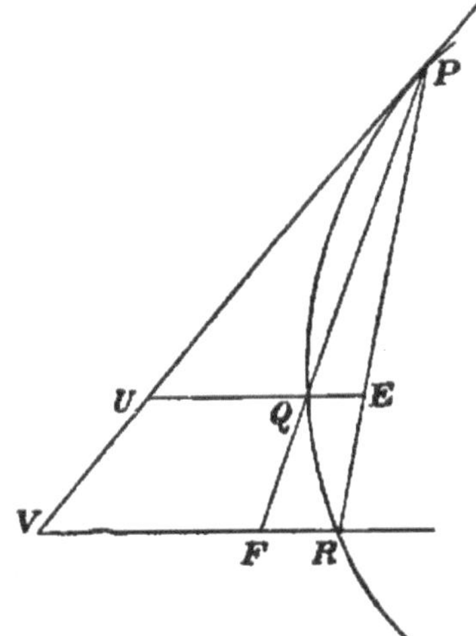

49. This relation may be presented in a different form, which is sometimes useful.

If from any point U in the tangent at P, UQ is drawn parallel to the axis, UP and UQ are respectively equal to the ordinate and abscissa of the point Q with regard to the diameter through P, and therefore

$$PU^2 = 4SP \,.\, UQ.$$

Therefore, if VR is drawn parallel to the axis from another point V of the tangent,

$$PU^2 : PV^2 :: UQ : VR.$$

Hence, since $$UE : VR :: PU : PV,$$

$$UE^2 : VR^2 :: UQ : VR :: UQ \,.\, VR : VR^2,$$

and $$UE^2 = UQ \,.\, VR.$$

Hence $$UE : UQ :: VR : UE :: PR : PE;$$

$$\therefore UQ : QE :: PE : ER.$$

In a similar manner it can be shewn that $VF^2 = UQ \,.\, VR$, and it follows that $VF = UE$, and therefore that EF is parallel to the tangent at P.

50. PROP. XX. *If QVQ' be a double ordinate of a diameter PV, and QD the perpendicular from Q upon PV, QD is a mean proportional between PV and the latus rectum.*

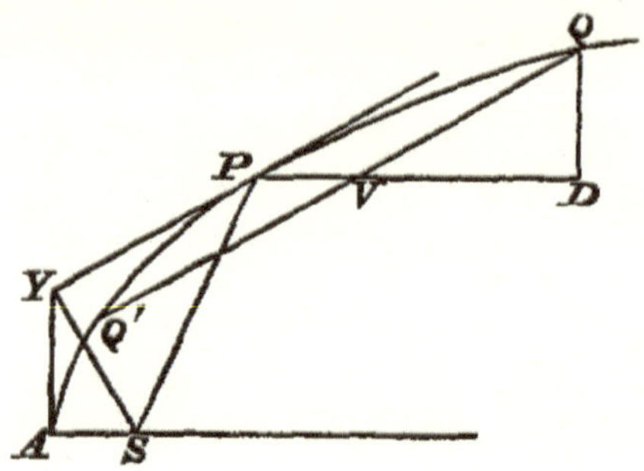

Let the tangent at P meet the tangent at the vertex in Y, and join SY.

The angle $QVD = SPY = SYA$, and therefore the triangles QVD, SAY are similar;

and
$$QD^2 : QV^2 :: AS^2 : SY^2$$
$$:: AS^2 : AS \,.\, SP.$$
$$:: AS : SP$$
$$:: 4AS \,.\, PV : 4SP \,.\, PV,$$

but $$QV^2 = 4SP \,.\, PV;$$

$$\therefore QD^2 = 4AS \,.\, PV.$$

51. PROP. XXI. *If from any point, within or without a parabola, two straight lines be drawn in given directions and intersecting the curve, the ratio of the rectangles of the segments is independent of the position of the point.*

From any point O draw a straight line intersecting the parabola in Q and Q', and draw the diameter OE, meeting the curve in E.

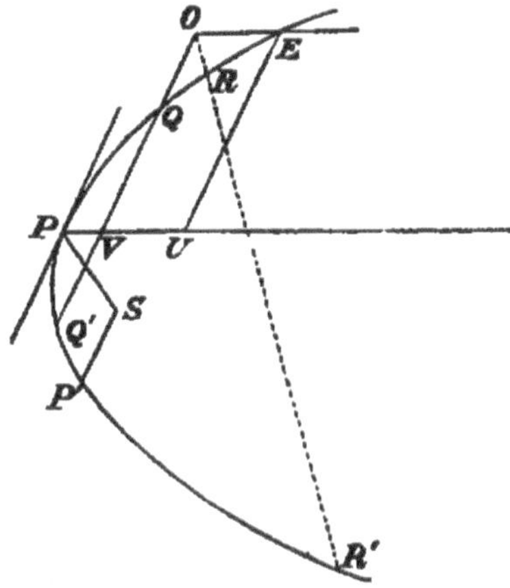

If PV be the diameter bisecting QQ', and EU the ordinate,

$$\begin{aligned} OQ\,.\,OQ' &= OV^2 - QV^2 \\ &= EU^2 - QV^2 = 4SP\,.\,PU - 4SP\,.\,PV \\ &= 4SP\,.\,OE. \end{aligned}$$

Similarly, if ORR' be any other intersecting line and P' the vertex of the diameter bisecting RR',

$$OR\,.\,OR' = 4SP'\,.\,OE.$$

$$\therefore OQ\,.\,OQ' : OR\,.\,OR' :: SP : SP',$$

that is, the ratio of the rectangles depends only on the positions of P and P', and, if the lines OQQ', ORR' are drawn parallel to given straight lines, these points P, P' are fixed.

It will be easily seen that the proof is the same if the point O be within the parabola.

If the lines OQQ', ORR' be moved parallel to themselves until they become the tangents at P and P', we shall then obtain, if these tangents intersect in T,

$$TP^2 : TP'^2 :: SP : SP';$$

a result previously obtained (Art. 38).

Again if QSQ', RSR' be the focal chords parallel to TP and TP', it follows that

$$TP^2 : TP'^2 :: QS\,.\,SQ' : RS\,.\,SR',$$

$\therefore$ (cor. Art. 8) $TP^2 : TP'^2 :: QQ' : RR'$.

52. PROP. XXII. *If from a point O, outside a parabola, a tangent OM, and a chord OAB be drawn, and if the diameter ME meet the chord in E,*

$$OE^2 = OA \,.\, OB.$$

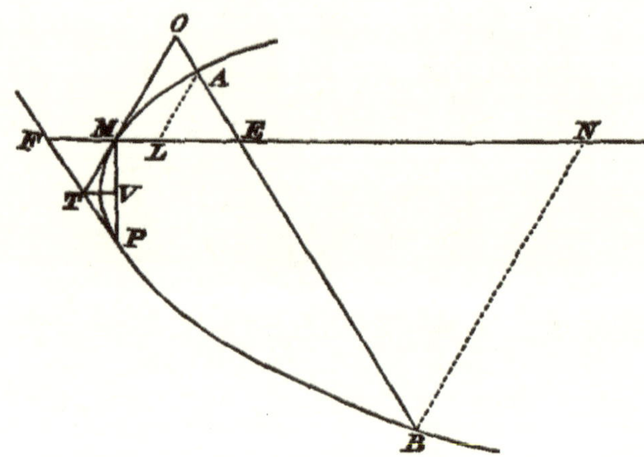

Let P be the point of contact of the tangent parallel to OAB, and let OM, ME meet this tangent in T and F.

Draw TV parallel to the axis and meeting PM in V;

then $$OA \,.\, OB : OM^2 :: TP^2 : TM^2 \text{ (Art. 51)},$$
$$:: TF^2 : TM^2,$$

since PM is bisected in V;

also $$TF : TM :: OE : OM;$$
$$\therefore OE^2 = OA \,.\, OB.$$

COR. 1. If AL, BN be the ordinates, parallel to OM, of A and B, ML, ME, and MN are proportional to OA, OE and OB, and therefore

$$ME^2 = ML \,.\, MN.$$

This theorem may be also stated in the following form:

If a chord AB of a parabola intersect a diameter in the point E, the distance of the point E from the tangent at the end of the diameter is a mean proportional between the distances of the points A and B from the same tangent.

COR. 2. Let KE be the ordinate through E parallel to OM.

Then, since $$ML : ME :: ME : MN,$$
$$AL^2 : KE^2 :: KE^2 : BN^2$$
$$\therefore AL : KE :: KE : BN,$$
so that KE is a mean proportional between AL and BN, the ordinates of A and B.

53. PROP. XXIII. *If a circle intersect a parabola in four points, the two straight lines constituting any one of the three pairs of the chords of intersection are equally inclined to the axis.*

Let Q, Q', R, R' be the four points of intersection;

then $$OQ \, . \, OQ' = OR \, . \, OR',$$
and therefore SP, SP' are equal, (Art. 51).

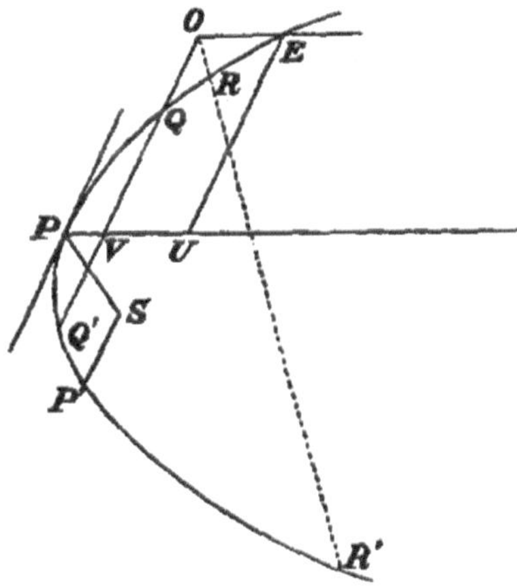

But, if SP, SP' be equal, the points P, P' are on opposite sides of, and are equidistant from the axis, and the tangents at P and P' are therefore equally inclined to the axis.

Hence the chords QQ', RR', which are parallel to these tangents, are equally inclined to the axis.

In the same manner it may be shewn that QR, $Q'R'$ are equally inclined to the axis, as also QR', $Q'R$.

54. Conversely, if two chords QQ', RR', which are not parallel, make equal angles with the axis, a circle can be drawn through Q, Q', R', R.

For, if the chords intersect in O, and OE be drawn parallel to the axis and meeting the curve in E, it may be shewn as above that

$$OQ \,.\, OQ' = 4SP \,.\, OE, \text{ and } OR \,.\, OR' = 4SP' \,.\, OE,$$

P and P' being the vertices of the diameters bisecting the chords.

But the tangents at P and P', which are parallel to the chords, are equally inclined to the axis, and therefore SP is equal to SP'.

Hence $$OQ \,.\, OQ' = OR \,.\, OR',$$

and therefore a circle can be drawn through the points Q, Q', R, R'.

If the two chords are both perpendicular to the axis, it is obvious that a circle can be drawn through their extremities, and this is the only case in which a circle can be drawn through the extremities of parallel chords.

EXAMPLES.

1. Find the locus of the centre of a circle which passes through a given point and touches a given straight line.

2. Draw a tangent to a parabola, making a given angle with the axis.

3. If the tangent at P meet the tangent at the vertex in Y,

$$AY^2 = AS \,.\, AN.$$

4. If the normal at P meet the axis in G, the focus is equidistant from the tangent at P and the straight line through G parallel to the tangent.

5. Given the focus, the position of the axis, and a tangent, construct the parabola.

6. Find the locus of the centre of a circle which touches a given straight line and a given circle.

7. Construct a parabola which has a given focus, and two given tangents.

8. The distance of any point on a parabola from the focus is equal to the length of the ordinate at that point produced to meet the tangent at the end of the latus rectum.

9. PT being the tangent at P, meeting the axis in T, and PN the ordinate, prove that $TY\,.\,TP = TS\,.\,TN$.

10. If SE be the perpendicular from the focus on the normal at P, shew that

$$SE^2 = AN\,.\,SP.$$

11. The locus of the vertices of all parabolas, which have a common focus and a common tangent, is a circle.

12. Having given the focus, the length of the latus rectum, and a tangent, construct the parabola.

13. If PSP' be a focal chord, and PN, $P'N'$ the ordinates, shew that

$$AN\,.\,AN' = AS^2.$$

Shew also that the latus rectum is a mean proportional between the double ordinates.

14. The locus of the middle points of the focal chords of a parabola is another parabola.

15. Shew that in general two parabolas can be drawn having a given straight line for directrix, and passing through two given points on the same side of the line.

16. Pp is a chord perpendicular to the axis, and the perpendicular from p on the tangent at P meets the diameter through P in R; prove that RP is equal to the latus rectum, and find the locus of R.

17. Having given the focus, describe a parabola passing through two given points.

18. The circle on any focal distance as diameter touches the tangent at the vertex.

19. The circle on any focal chord as diameter touches the directrix.

20. A point moves so that its shortest distance from a given circle is equal to its distance from a given diameter of the circle; prove that the locus is a parabola, the focus of which coincides with the centre of the circle.

21. Find the locus of a point which moves so that its shortest distance from a given circle is equal to its distance from a given straight line.

22. The vertex of an isosceles triangle is fixed. The extremities of its base lie on two fixed parallel straight lines. Prove that the base is a tangent to a parabola.

23. Shew that the normal at any point of a parabola is equal to the ordinate through the middle point of the subnormal.

24. If perpendiculars are drawn to the tangents to a parabola where they meet the axis they will be normals to two equal parabolas.

25. PSP' is a focal chord of a parabola. The diameters through P, P' meet the normals at P', P in V, V' respectively. Prove that $PVV'P'$ is a parallelogram.

26. If APC be a sector of a circle, of which the radius CA is fixed, and a circle be described, touching the radii CA, CP, and the arc AP, the locus of the centre of this circle is a parabola.

27. If from the focus S of a parabola, SY, SZ be perpendiculars drawn to the tangent and normal at any point, YZ is parallel to the diameter.

28. Prove that the locus of the foot of the perpendicular from the focus on the normal is a parabola.

29. If PG be the normal, and GL the perpendicular from G upon SP, prove that GL is equal to the ordinate PN.

30. Given the focus, a point P on the curve, and the length of the perpendicular from the focus on the tangent at P, find the vertex.

31. A circle is described on the latus rectum as diameter, and a common tangent QP is drawn to it and the parabola: shew that SP, SQ make equal angles with the latus rectum.

32. G is the foot of the normal at a point P of the parabola, Q is the middle point of SG, and X is the foot of the directrix: prove that

$$QX^2 - QP^2 = 4AS^2.$$

33. If PG the normal at P meet the axis in G, and if PF, PH, lines equally inclined to PG, meet the axis in F and H, the length SG is a mean proportional between SF and SH.

34. A triangle ABC circumscribes a parabola whose focus is S, and through A, B, C, lines are drawn respectively perpendicular to SA, SB, SC; shew that these pass through one point.

35. If PQ be the normal at P meeting the curve in Q, and if the chord PR be drawn so that PR, PQ are equally inclined to the axis, PRQ is a right angle.

36. PN is a semi-ordinate of a parabola, and AM is taken on the other side of the vertex along the axis equal to AN; from any point Q in PN, QR is drawn parallel to the axis meeting the curve in R; prove that the lines MR, AQ will intersect in the parabola.

37. Having given two points of a parabola, the direction of the axis, and the tangent at one of the points, construct the parabola.

38. Having given the vertex of a diameter, and a corresponding double ordinate, construct the parabola.

39. PM is an ordinate of a point P; a straight line parallel to the axis bisects PM, and meets the curve in Q; MQ meets the tangent at the vertex in T; prove that $3AT = 2PM$.

40. AB, CD are two parallel straight lines given in position, and AC is perpendicular to both, A and C being given points; in CD any point Q is taken, and in AQ, produced if necessary, a point P is taken, such that the distance of P from AB is equal to CQ; prove that the locus of P is a parabola.

41. If the tangent and normal at a point P of a parabola meet the tangent at the vertex in K and L respectively, prove that

$$KL^2 : SP^2 :: SP - AS : AS.$$

42. Having given the length of a focal chord, find its position.

43. If the ordinate of a point P bisects the subnormal of a point P', prove that the ordinate of P is equal to the normal of P'.

44. A parabola being traced on a plane, find its axis and vertex.

45. If PV, $P'V'$ be two diameters, and PV', $P'V$ ordinates to these diameters,

$$PV = P'V'.$$

46. If one side of a triangle be parallel to the axis of a parabola, the other sides will be in the ratio of the tangents parallel to them.

47. QVQ' is an ordinate of a diameter PV, and any chord PR meets QQ' in N, and the diameter through Q in L; prove that

$$PL^2 = PN \, . \, PR.$$

48. Describe a parabola passing through three given points, and having its axis parallel to a given line.

49. If AP, AQ be two chords drawn from the vertex at right angles to each other, and PN, QM be ordinates, the latus rectum is a mean proportional between AN and AM.

50. PSp is a focal chord of a parabola; prove that AP, Ap meet the latus rectum in two points whose distances from the focus are equal to the ordinates of p and P respectively.

51. If the straight line AP and the diameter through P meet the double ordinate QMQ' in R and R', prove that

$$RM \, . \, R'M = QM^2.$$

52. A and P are two fixed points. Parabolas are drawn all having their vertices at A, and all passing through P. Prove that the points of intersection of the tangents at P with the tangent and normal at A lie on two fixed circles, one of which is double the size of the other.

53. A variable tangent to a parabola intersects two fixed tangents in the points T and T': shew that the ratio $ST : ST'$ is constant.

54. Through a fixed point on the axis of a parabola a chord PQ is drawn, and a circle of given radius is described through the feet of the ordinates of P and Q. Shew that the locus of its centre is a circle.

55. If SY be the perpendicular on the tangent at P, and if YS be produced to R so that $SR = SY$, shew that PAR is a right angle.

56. If two circles be drawn touching a parabola at the ends of a focal chord, and passing through the focus, shew that they intersect each other orthogonally.

57. PSQ is a focal chord of a parabola, whose vertex is A and focus S, V being the middle point of the chord, shew that

$$PV^2 = AV^2 + 3AS^2.$$

58. QQ' is a focal chord of a parabola. Describe a circle which shall pass through Q, Q' and touch the parabola.

If P be the point of contact and the angle QPQ' a right angle, find the inclination of QP to the axis.

59. Through two fixed points E, F, on the axis of a parabola are drawn two chords PQ, PR meeting the curve in P, Q, R. If QR meet the axis in T, shew that the ratio $TR : TQ$ is constant.

60. A chord PQ is normal to the parabola at P, and the angle PSQ is a right angle. Prove that $SQ = 2SP$, and that the ordinate of P is equal to the latus rectum. Also, if T is the point of intersection of the tangents at P and Q, and if R is the middle point of TQ, prove that the angle TSR is a right angle, and that $ST = 2SR$.

61. A straight line intersects a circle; prove that all the chords of the circle which are bisected by the straight line are tangents to a parabola.

62. If two tangents TP, TQ be drawn to a parabola, the perpendicular SE from the focus on their chord of contact passes through the middle point of their intercept on the tangent at the vertex.

63. From the vertex of a parabola a perpendicular is drawn on the tangent at any point; prove that the locus of its intersection with the diameter through the point is a straight line.

64. If two tangents to a parabola be drawn from any point in its axis, and if any other tangent intersect these two in P and Q, prove that $SP = SQ$.

65. T is a point on the tangent at P, such that the perpendicular from T on SP is of constant length; prove that the locus of T is a parabola.

If the constant length be $2AS$, prove that the vertex of the locus is on the directrix.

66. Given a chord of a parabola in magnitude and position, and the point in which the axis cuts the chord, the locus of the vertex is a circle.

67. If the normal at a point P of a parabola meet the curve in Q, and the tangents at P and Q intersect in T, prove that T and P are equidistant from the directrix.

68. If TP, TQ be tangents to a parabola, such that the chord PQ is normal at P,

$$PQ : PT :: PN : AN,$$

PN and AN being the ordinate and abscissa.

69. If two equal tangents to a parabola be cut by a third tangent, the alternate segments of the two tangents will be equal.

70. If AP be a chord through the vertex, and if PL, perpendicular to AP, and PG, the normal at P, meet the axis in L, G respectively, GL = half the latus rectum.

71. If PSQ be a focal chord, A the vertex, and PA, QA be produced to meet the directrix in P', Q' respectively, then $P'SQ'$ will be a right angle.

72. The tangents at P and Q intersect in T, and the tangent at R intersects TP and TQ in C and D; prove that

$$PC : CT :: CR : RD :: TD : DQ.$$

73. From any point D in the latus rectum of a parabola, a straight line DP is drawn, parallel to the axis, to meet the curve in P; if X be the foot of the directrix, and A the vertex, prove that AD, XP intersect in the parabola.

74. PSp is a focal chord, and upon PS and pS as diameters circles are described; prove that the length of either of their common tangents is a mean proportional between AS and Pp.

75. If AQ be a chord of a parabola through the vertex A, and QR be drawn perpendicular to AQ to meet the axis in R; prove that AR will be equal to the chord through the focus parallel to AQ.

76. If from any point P of a circle, PC be drawn to the centre C, and a chord PQ be drawn parallel to the diameter AB, and bisected in R; shew that the locus of the intersection of CP and AR is a parabola.

77. A circle, the diameter of which is three-fourths of the latus rectum, is described about the vertex A of a parabola as centre; prove that the common chord bisects AS.

78. Shew that straight lines drawn perpendicular to the tangents of a parabola through the points where they meet a given fixed line perpendicular to the axis are in general tangents to a confocal parabola.

79. If QR be a double ordinate, and PD a straight line drawn parallel to the axis from any point P of the curve, and meeting QR in D, prove, from Art. 27, that

$$QD \,.\, RD = 4AS \,.\, PD.$$

80. Prove, by help of the preceding theorem, that, if QQ' be a chord parallel to the tangent at P, QQ' is bisected by PD, and hence determine the locus of the middle point of a series of parallel chords.

81. If a parabola touch the sides of an equilateral triangle, the focal distance of any vertex of the triangle passes through the point of contact of the opposite side.

82. Find the locus of the foci of the parabolas which have a common vertex and a common tangent.

83. From the points where the normals to a parabola meet the axis, lines are drawn perpendicular to the normals: shew that these lines will be tangents to an equal parabola.

84. Inscribe in a given parabola a triangle having its sides parallel to three given straight lines.

85. PNP' is a double ordinate, and through a point of the parabola RQL is drawn perpendicular to PP' and meeting PA, or PA produced in R; prove that

$$PN : NL :: LR : RQ.$$

86. PNP' is a double ordinate, and through R, a point in the tangent at P, RQM is drawn perpendicular to PP' and meeting the curve in Q; prove that

$$QM : QR :: P'M : PM.$$

87. If from the point of contact of a tangent to a parabola, a chord be drawn, and a line parallel to the axis meeting the chord, the tangent, and the curve, shew that this line will be divided by them in the same ratio as it divides the chord.

88. PSp is a focal chord of a parabola, RD is the directrix meeting the axis in D, Q is any point in the curve; prove that if QP, Qp produced meet the directrix in R, r, half the latus rectum will be a mean proportional between DR and Dr.

89. A chord of a parabola is drawn parallel to a given straight line, and on this chord as diameter a circle is described; prove that the distance between the middle points of this chord, and of the chord joining the other two points of intersection of the circle and parabola, will be of constant length.

90. If a circle and a parabola have a common tangent at P, and intersect in Q and R; and if QV, UR be drawn parallel to the axis of the parabola meeting the circle in V and U respectively, then will VU be parallel to the tangent at P.

91. If PV be the diameter through any point P, QV a semi-ordinate, Q' another point in the curve, and $Q'P$ cut QV in R, and $Q'R'$, the diameter through Q', meet QV in R', then

$$VR\,.\,VR' = QV^2.$$

92. PQ, PR are any two chords; PQ meets the diameter through R in the point F, and PR meets the diameter through Q in E; prove that EF is parallel to the tangent at P.

93. If parallel chords be intersected by a diameter, the distances of the points of intersection from the vertex of the diameter are in the ratio of the rectangles contained by the segments of the chords.

94. If tangents be drawn to a parabola from any point P in the latus rectum, and if Q, Q' be the points of contact, the semi-latus rectum is a geometric mean between the ordinates of Q and Q', and the distance of P from the axis is an arithmetic mean between the same ordinates.

95. If A', B', C' be the middle points of the sides of a triangle ABC, and a parabola drawn through A', B', C' meet the sides again in A'', B'', C'', then will the lines AA'', BB'', CC'' be parallel to each other.

96. A circle passing through the focus cuts the parabola in two points. Prove that the angle between the tangents to the circle at those points is four times the angle between the tangents to the parabola at the same points.

97. The locus of the points of intersection of normals at the extremities of focal chords of a parabola is another parabola.

98. Having given the vertex, a tangent, and its point of contact, construct the parabola.

99. PSp is a focal chord of a parabola; shew that the distance of the point of intersection of the normals at P and p from the directrix varies as the rectangle contained by PS, pS.

100. TP, TQ are tangents to a parabola at P and Q, and O is the centre of the circle circumscribing PTQ; prove that TSO is a right angle.

101. P is any point of a parabola whose vertex is A, and through the focus S the chord QSQ' is drawn parallel to AP; PN, QM, $Q'M'$, being perpendicular to the axis, shew that SM is a mean proportional between AM, AN, and that

$$MM' = AP.$$

102. If a circle cut a parabola in four points, two on one side of the axis, and two on the other, the sum of the ordinates of the first two is equal to the sum of the ordinates of the other two points.

Extend this theorem to the case in which three of the points are on one side of the axis and one on the other.

103. The tangents at P and Q meet in T, and TL is the perpendicular from T on the axis; prove that if PN, QM be the ordinates of P and Q,

$$PN \, . \, QM = 4AS \, . \, AL.$$

104. The tangents at P and Q meet in T, and the lines TA, PA, QA, meet the directrix in t, p, and q: prove that

$$tp = tq.$$

105. From a point T tangents TP, TQ are drawn to a parabola, and through T straight lines are drawn parallel to the normals at P and Q; prove that one diagonal of the parallelogram so formed passes through the focus.

106. Through a given point within a parabola draw a chord which shall be divided in a given ratio at that point.

107. ABC is a portion of a parabola bounded by the axis AB and the semi-ordinate BC; find the point P in the semi-ordinate such that if PQ be drawn parallel to the axis to meet the parabola in Q, the sum of BP and PQ shall be the greatest possible.

108. The diameter through a point P of a parabola meets the tangent at the vertex in Z; the normal at P and the focal distance of Z will intersect in a point at the same distance from the tangent at the vertex as P.

109. Given a tangent to a parabola and a point on the curve, shew that the foot of the ordinate of the point of contact of the tangent drawn to the diameter through the given point lies on a fixed straight line.

110. Find a point such that the tangents from it to a parabola and the lines from the focus to the points of contact may form a parallelogram.

111. Two equal parabolas have a common focus; and, from any point in the common tangent, another tangent is drawn to each; prove that these tangents are equidistant from the common focus.

112. Two parabolas have a common axis and vertex, and their concavities turned in opposite directions; the latus rectum of one is eight times that of the other; prove that the portion of a tangent to the former, intercepted between the common tangent and axis, is bisected by the latter.

CHAPTER III.

THE ELLIPSE.

DEF. *An ellipse is the curve traced out by a point which moves in such a manner that its distance from a given point is in a constant ratio of less inequality to its distance from a given straight line.*

Tracing the Curve.

55. Let S be the focus, EX the directrix, and SX the perpendicular on EX from S.

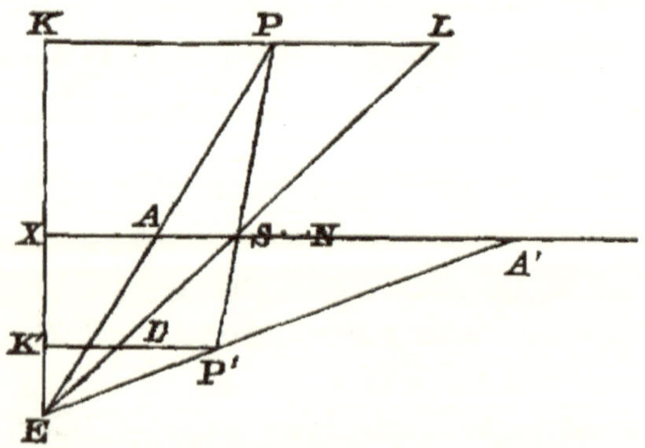

Divide SX at the point A in the given ratio; the point A is the vertex.

From any point E in EX, draw EAP, ESL, and through S draw SP making the angle PSL equal to LSN, and meeting EAP in P.

Through P draw LPK perpendicular to the directrix and meeting ESL in L.

Then the angle $\qquad PSL = LSN = SLP.$

$$\therefore SP = PL.$$

Also $\qquad PL : PK :: SA : AX.$

Hence $\qquad SP : PK :: SA : AX,$

and P is therefore a point in the curve.

Again, in the axis XAN find a point A' such that

$$SA' : A'X :: SA : AX;$$

this point is evidently on the same side of the directrix as the point A, and is another vertex of the curve.

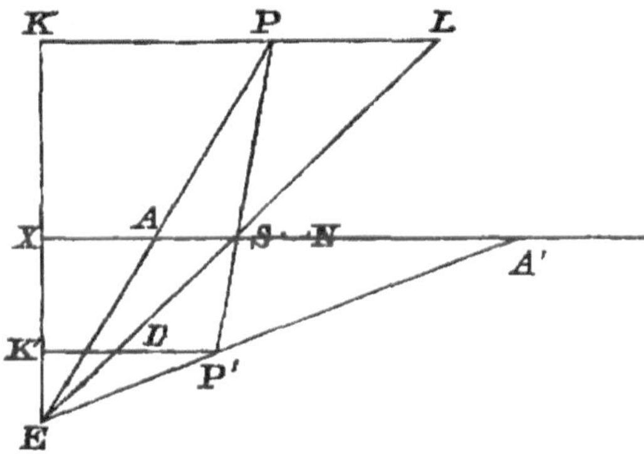

Join EA' meeting PS produced in P', and draw $P'L'K'$ perpendicular to the directrix and meeting ES in L'.

Then $\qquad P'L' : P'K' :: SA' : A'X$

$$:: SA : AX,$$

and the angle $\qquad SL'P' = L'SA = L'SP;$

$$\therefore P'L' = SP'.$$

Hence P' is also a point in the curve, and PSP' is a focal chord.

By giving E a series of positions on the directrix we shall obtain a series of focal chords, and we can also, as in Art. (1), find other points of the curve lying in the lines KP, $K'P'$, or in these lines produced.

We can thus find any number of points in the curve.

56. DEF. *The distance AA' is the major axis.*

The middle point C of AA' is called the centre of the ellipse.

If through C the double ordinate BCB' be drawn, BB' is called the minor axis.

Any straight line drawn through the centre, and terminated by the curve, is called a diameter.

The lines ACA', BCB' are called the principal diameters, or, briefly, the axes of the curve.

The line ACA' is also sometimes called the transverse axis, and BCB' the conjugate axis.

57. PROP. I. *If P be any point of an ellipse, and AA' the axis major, and if PA, $A'P$, when produced, meet the directrix in E and F, the distance EF subtends a right angle at the focus.*

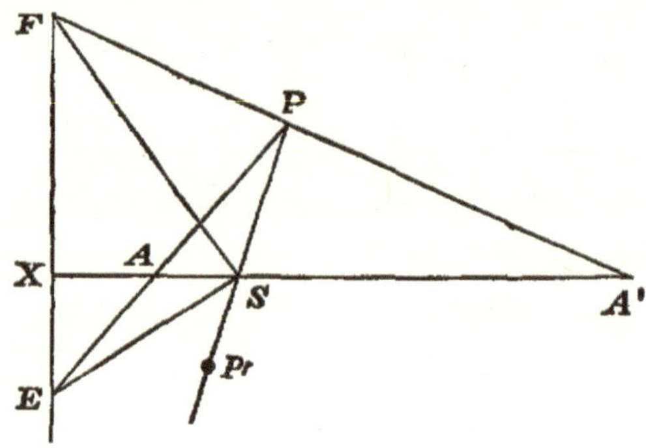

By the theorem of Art. 4, ES bisects the angle ASP', and FS bisects the angle ASP;

$$\therefore ESF \text{ is a right angle.}$$

It will be seen that, since ASA' is a focal chord, this is a particular case of the theorem of Art. 6.

58. PROP. II. *If PN be the ordinate of any point P of an ellipse, ACA' the axis major, and BCB' the axis minor,*

$$PN^2 : AN \,.\, NA' :: BC^2 : AC^2.$$

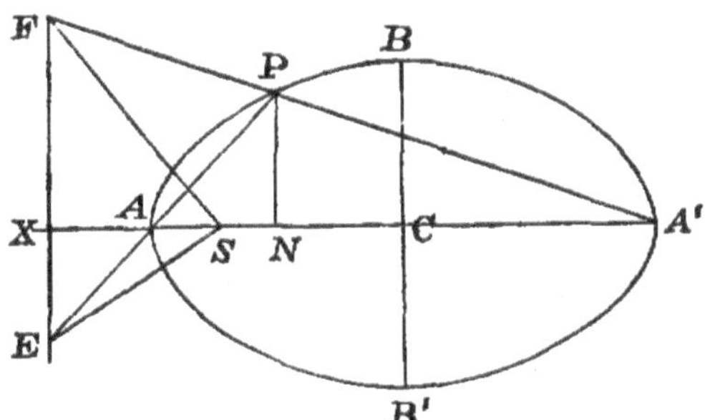

Join PA, $A'P$, and let these lines produced meet the directrix in E and F.

Then $$PN : AN :: EX : AX,$$

and $$PN : A'N :: FX : A'X;$$

$$\therefore PN^2 : AN \,.\, NA' :: EX \,.\, FX : AX \,.\, A'X$$
$$:: SX^2 : AX \,.\, A'X,$$

since ESF is a right angle (Prop. 1.); that is, PN^2 is to $AN \,.\, NA'$ in a constant ratio.

Hence, taking PN coincident with BC, in which case

$$AN = NA' = AC,$$
$$BC^2 : AC^2 :: SX^2 : AX \,.\, A'X,$$

and $$\therefore PN^2 : AN \,.\, NA' :: BC^2 : AC^2.$$

This may be also written

$$PN^2 : AC^2 - CN^2 :: BC^2 : AC^2.$$

Cor. If PM be the perpendicular from P on the axis minor,

$$CM = PN, PM = CN,$$

and $$CM^2 : AC^2 - PM^2 :: BC^2 : AC^2.$$

Hence $$AC^2 : AC^2 - PM^2 :: BC^2 : CM^2,$$

and $$\therefore AC^2 : PM^2 :: BC^2 : BC^2 - CM^2,$$

or $$PM^2 : BM \,.\, MB' :: AC^2 : BC^2.$$

59. If a point N' be taken on the axis major, between C and A', such that $CN' = CN$, the corresponding ordinate $P'N' = PN$, and therefore it follows that the curve is symmetrical with regard to BCB', and that there is another focus, and another directrix, corresponding to the vertex A'.

60. By help of the theorem of Art. 57, we can give an independent proof of the existence of the other focus and directrix, corresponding to the vertex A'.

In AA' produced take a point X' such that $A'X' = AX$, and in AA' take a point S' such that $A'S' = AS$.

Through X' draw a straight line $eX'f$ perpendicular to the axis, and let EP, FP produced meet this line in e and f. Join eS', and fS'.

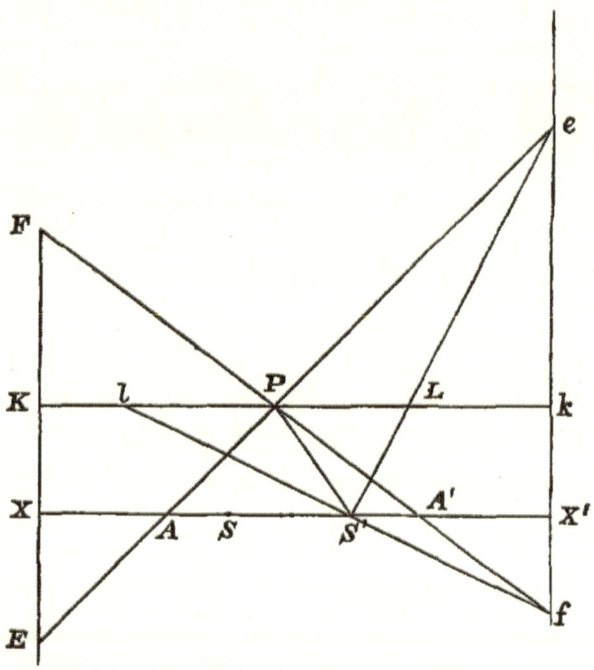

Then
$$eX' : EX :: AX' : AX$$
$$:: A'X : A'X'$$
$$:: FX : fX';$$
$$\therefore eX' \,.\, fX' = EX \,.\, FX = SX^2 = S'X'^2.$$

Hence $eS'f$ is a right angle.

Through P draw KPk parallel to the axis, meeting eS' and fS' produced in L and l.

Then $$PL : Pk :: S'A : AX' :: SA' : A'X,$$

and $$Pl : Pk :: S'A' : A'X' :: SA : AX,$$

$$\therefore PL = Pl.$$

Moreover, $LS'l$ being a right angle,

$$S'P = Pl,$$

$$\therefore S'P : Pk :: S'A' : A'X',$$

and the curve can be described by means of the focus S' and the directrix eX'.

If SA be equal to AX, the point A', and therefore the points S' and X', will be at an infinite distance from S and A.

Hence a parabola is the limiting form of an ellipse, the axis major of which is indefinitely increased in magnitude, while the distance SA remains finite.

61. Prop. III. *If ACA' be the axis major, C the centre, S one of the foci, and X the foot of the directrix,*

$$CS : CA :: CA : CX :: SA : AX,$$

and $$CS : CX :: CS^2 : CA^2.$$

X A S C S' A' X'

For $$S'A : SA :: AX' : AX$$

$$:: A'X : AX;$$

$$\therefore SS' : SA :: AA' : AX,$$

or $$CS : CA :: SA : AX.$$

Again, $$SA' : SA :: AX' : AX;$$

$$\therefore AA' : SA :: XX' : AX,$$

or $$CA : CX :: SA : AX;$$

$$\therefore CS : CA :: CA : CX,$$

or $$CS . CX = CA^2.$$

Also $$CS : CX :: CS^2 : CS . CX$$

$$:: CS^2 : CA^2.$$

62. PROP. IV. *If S be a focus, and B an extremity of the axis minor,*

$$SB = AC \text{ and } BC^2 = AS \,.\, SA'.$$

For, joining SB in the figure of Art. 58,

$$SB : CX :: SA : AX$$
$$:: CA : CX,$$

by the previous Article,

$$\therefore SB = CA.$$

Also $$BC^2 = SB^2 - SC^2 = AC^2 - SC^2$$
$$= AS \,.\, SA'.$$

63. PROP. V. *The semi-latus rectum SR is a third proportional to AC and BC.*

For, Prop. II.,

$$SR^2 : AS \,.\, SA' :: BC^2 : AC^2;$$
$$\therefore SR^2 : BC^2 :: BC^2 : AC^2,$$

or $$SR : BC :: BC : AC.$$

COR. Since $$SR : SX :: SA : AX$$
$$:: SC : AC,$$

it follows that $SX \,.\, SC = SR \,.\, AC = BC^2$;
and hence also, since $SC \,.\, CX = AC^2$, that

$$SX : CX :: BC^2 : AC^2.$$

64. PROP. VI. *The sum of the focal distances of any point is equal to the axis major.*

Let PN be the ordinate of a point P (Fig. Art. 60), then

$$S'P : SP :: NX' : NX;$$
$$\therefore S'P + SP : SP :: XX' : NX,$$

or $$S'P + SP : XX' :: SP : NX$$
$$:: SA : AX$$
$$:: AA' : XX';$$
$$\therefore S'P + SP = AA'$$

COR. Since $SP : NX :: SA : AX$

$:: AC : CX;$

$\therefore AC : SP :: CX : NX,$

$AC - SP : SP :: CN : NX,$

and $AC - SP : CN :: SA : AX.$

Also, $AC - SP = S'P - AC;$

$\therefore S'P - AC : CN :: SA : AX.$

Hence, $S'P - SP : 2CN :: SA : AX.$

Mechanical Construction of the Ellipse.

65. Fasten the ends of a piece of thread to two pins fixed on a board, and trace a curve on the board with a pencil pressed against the thread so as to keep it stretched; the curve traced out will be an ellipse, having its foci at the points where the pins are fixed, and having its major axis equal to the length of the thread.

66. PROP. VII. *The sum of the distances of a point from the foci of an ellipse is greater or less than the major axis according as the point is outside or inside the ellipse.*

If the point be without the ellipse, join SQ, $S'Q$, and take a point P on the intercepted arc of the curve.

Then P is within the triangle SQS' and therefore, joining SP, $S'P$,

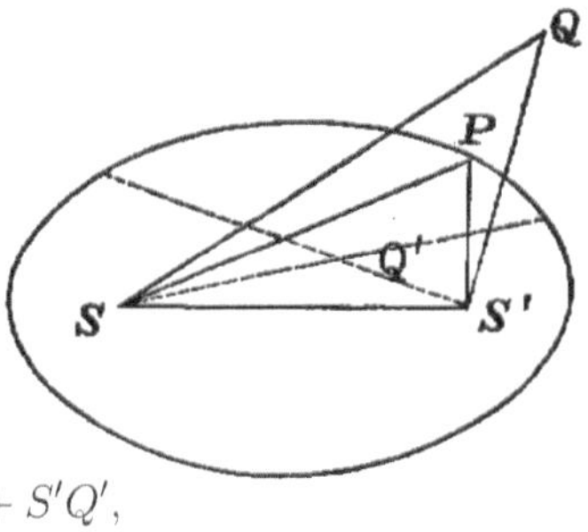

$SQ + S'Q > SP + S'P$, Euclid I. 21,

i.e. $SQ + S'Q > AA'$.

If Q' be within the ellipse, let SQ', $S'Q'$ produced meet the curve and take a point P on the intercepted arc.

Then Q' is within the triangle SPS', and

$$\therefore SP + S'P > SQ' + S'Q',$$

i.e. $$SQ' + S'Q' < AA'.$$

67. DEF. *The circle described on the axis major as diameter is called the auxiliary circle.*

PROP. VIII. *If the ordinate NP of an ellipse be produced to meet the auxiliary circle in Q,*

$$PN : QN :: BC : AC.$$

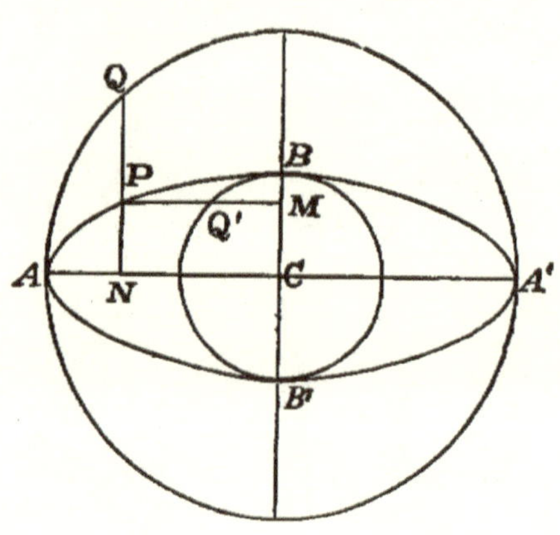

For (Art. 58)

$$PN^2 : AN \,.\, NA' :: BC^2 : AC^2,$$

and, by a property of the circle,

$$QN^2 = AN \,.\, NA'; \therefore PN : QN :: BC : AC.$$

COR. Similarly, if PM, the perpendicular on BB', meet in Q' the circle described on BB' as diameter,

$$PM : Q'M :: AC : BC.$$

For $$PM^2 : BM \,.\, MB' :: AC^2 : BC^2,$$

and $$BM \,.\, MB' = Q'M^2.$$

Properties of the Tangent and Normal.

68. PROP. IX. *The normal at any point bisects the angle between the focal distances of that point, and the tangent is equally inclined to the focal distances.*

Let the normal at P meet the axis in G; then (Art. 18)

$$SG : SP :: SA : AX,$$

and $$S'G : S'P :: SA : AX.$$

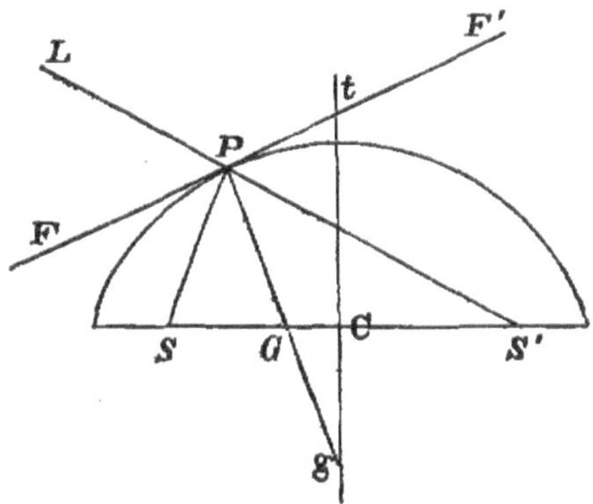

Hence $$SG : S'G :: SP : S'P,$$

and therefore the angle SPS' is bisected by PG.

Also FPF' being the tangent, and GPF, GPF' being right angles, it follows that the angles SPF, $S'PF'$ are equal, or that the tangent is equally inclined to the focal distances.

Hence if $S'P$ be produced to L, the tangent bisects the angle SPL.

Cor. If a circle be described about the triangle SPS', its centre will lie in BCB', which bisects SS' at right angles; and since the angles SPG, $S'PG$ are equal, and equal angles stand upon equal arcs, the point g, in which PG produced meets the minor axis, is a point in the circle.

Also, if the tangent meet the minor axis in t, the point t is on the same circle, since gPt is a right angle.

Hence, *Any point P of an ellipse, the two foci, and the points of intersection of the tangent and normal at P with the minor axis are concyclic.*

69. Prop. X. *Every diameter is bisected at the centre, and the tangents at the ends of a diameter are parallel.*

Let PCp be a diameter, PN, pn the ordinates of P and p.

Then $$CN^2 : Cn^2 :: PN^2 : pn^2$$
$$:: AC^2 - CN^2 : AC^2 - Cn^2 \text{ (Art. 58)};$$
$$\therefore CN^2 : AC^2 :: Cn^2 : AC^2.$$

Hence $$CN = Cn \text{ and } \therefore CP = Cp.$$

Draw the focal distances; then, since Pp and SS' bisect each other in C, the figure $SPS'p$ is a parallelogram, and the angle

$$SPS' = SpS'.$$

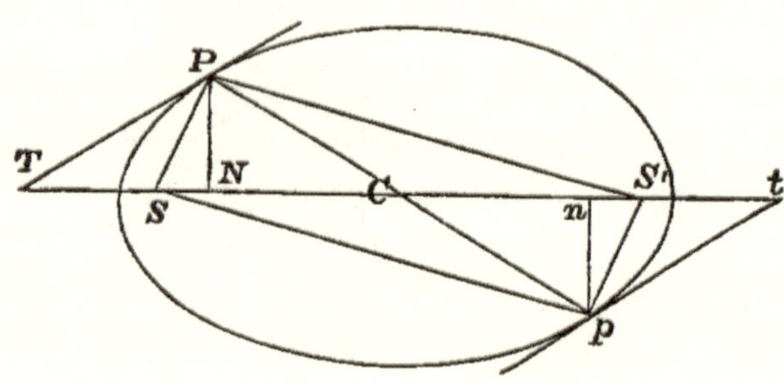

But the tangents PT, pt are equally inclined to the focal distances;

$$\therefore \text{ the angle } SPT = S'pt,$$

and, adding the equal angles CPS, CpS',

$$CPT = Cpt;$$

$$\therefore PT \text{ and } pt \text{ are parallel.}$$

COR. Since Sp and $S'p$ are equally inclined to the tangent at p, it follows that SP and Sp make equal angles with the tangents at P and p.

70. PROP. XI. *The perpendiculars from the foci on any tangent meet the tangent on the auxiliary circle, and the semi-minor axis is a mean proportional between their lengths.*

Let SY, $S'Y'$ be the perpendiculars; join $S'P$, and let SY, $S'P$ produced meet in L.

The angles SPY, YPL being equal, and PY being common, the triangles SPY, YPL are equal in all respects;

$$\therefore PL = SP,\ SY = YL,$$

and $$S'L = S'P + PL = S'P + SP = AA'.$$

Join CY, then C being the middle point of SS', and Y of SL, CY is parallel to $S'L$,

and $$\therefore S'L = 2CY.$$

Hence $CY = AC$, and Y is a point on the auxiliary circle.

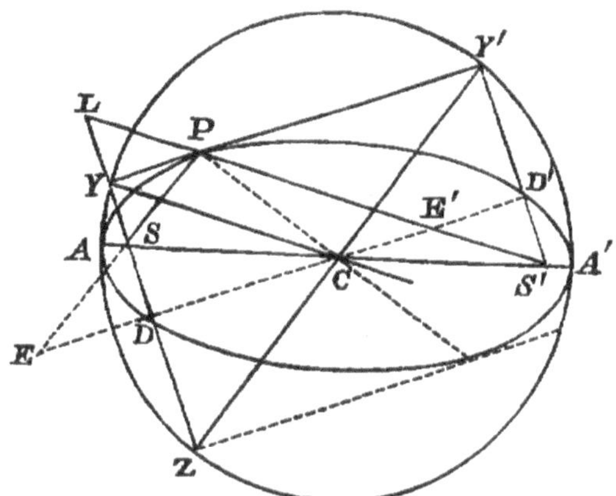

Similarly by producing SP, $S'Y'$ it may be shewn that Y' is also on the auxiliary circle.

Let YS produced meet the circle in Z, and join $Y'Z$; then $Y'YZ$ being a right angle, $Y'Z$ is a diameter and passes through C.

Hence the triangles SCZ, $S'CY'$ are equal, and

$$SY \,.\, S'Y' = SY \,.\, SZ = AS \,.\, SA' = BC^2.$$

COR. (1). If P' be the other extremity of the diameter through P, the tangent at P' is parallel to PY, and therefore Z is the foot of the perpendicular from S on the tangent at P'.

COR. (2). If the diameter DCD', drawn parallel to the tangent at P, meet SP, $S'P$ in E and E', $PECY'$ is a parallelogram, for CY' is parallel to SP, and CE to PY';

$$\therefore PE = CY' = AC; \text{ and similarly } PE' = CY = AC.$$

COR. (3). Any diameter parallel to the focal distance of a point meets the tangent at the point on the auxiliary circle.

71. PROP. XII. *To draw tangents from a given point to an ellipse.*

For this purpose we may employ the general construction of Art. (17), or the following.

Let Q be the given point; upon SQ as diameter describe a circle cutting the auxiliary circle in Y and Y'; YQ and $Y'Q$ will be the required tangents.

Producing SY to L so that $YL = SY$, join $S'L$ cutting the line YQ in P.

The triangles SPY, LPY are equal in all respects, since $SY = YL$ and PY is common and perpendicular to SL;

$$\therefore SP = PL, \text{ and } S'L = S'P + PL = S'P + SP;$$

but, joining CY, $S'L = 2CY = 2AC$;

$$\therefore SP + S'P = 2AC,$$

and P is therefore a point on the ellipse.

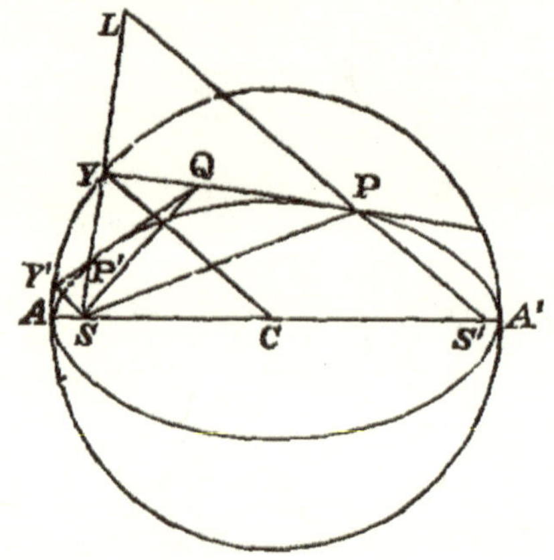

Also the angle $SPY = YPL$,

and $\therefore QP$ is the tangent at P.

A similar construction will give the point of contact of the other tangent QP'.

Referring to Art. 35 it will be seen that the construction is the same as that given for the parabola, the ultimate form of the circle being, for the parabola, the tangent at the vertex.

72. Prop. XIII. *If two tangents be drawn to an ellipse from an external point, they are equally inclined to the focal distances of that point.*

Let QP, QP' be the tangents, SY, $S'Y'$, SZ, $S'Z'$ the perpendiculars from the foci on the tangents; join YZ, $Y'Z'$.

Then (Art. 70)

$$SY \,.\, S'Y' = SZ \,.\, S'Z';$$
$$\therefore SY : SZ :: S'Z' : S'Y'.$$

The points S, Y, Q, Z being concyclic, the angles YSZ, YQZ are supplementary; and similarly, $Z'S'Y'$, $Z'QY'$ are supplementary.

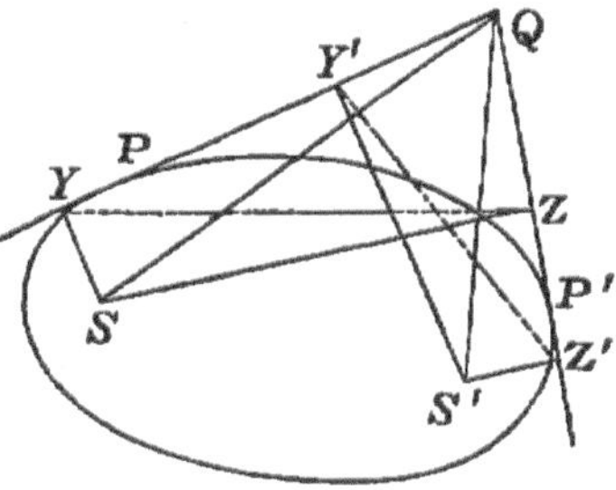

Therefore the angle $YSZ = Z'S'Y'$ and the triangles YSZ, $Z'S'Y'$ are similar.

Therefore the angle $SQP = SZY = S'Y'Z' = S'QP'$.

73. DEF. *Ellipses which have the same foci are called confocal ellipses.*

If Q be a point in a confocal ellipse the normal at Q bisects the angle SQS' and therefore bisects the angle PQP'.

Hence, *If from any point of an ellipse tangents are drawn to a confocal ellipse, these tangents are equally inclined to the normal at the point.*

By reference to the remark of Art. 41, it will be seen that this theorem includes that of Art. 41 as a particular case.

74. PROP. XIV. If PT the tangent at P meet the axis major in T, and PN be the ordinate,

$$CN \,.\, CT = AC^2.$$

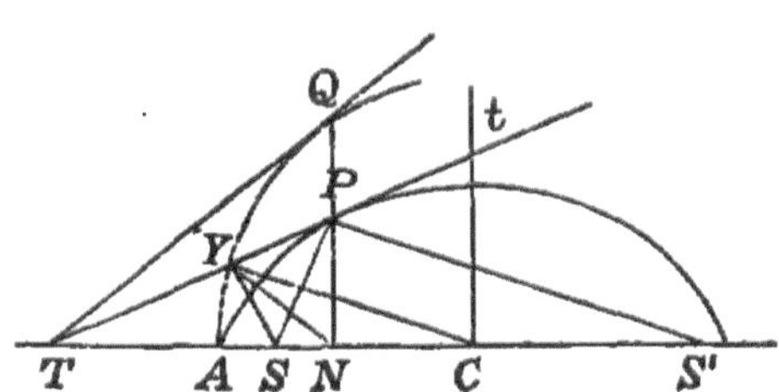

Draw the focal distances SP, $S'P$, and the perpendicular SY on the tangent, and join NY, CY.

Then, as in Art. 70, CY is parallel to $S'P$; therefore the angle

$$CYP = S'Pt = SPY$$
$$= SNY,$$

since S, Y, P, N are concyclic.

Hence $$CYT = CNY,$$

and the triangles CYT, CNY are equiangular.

Therefore $$CN : CY :: CY : CT$$
or $$CN \,.\, CT = CY^2 = AC^2.$$

COR. (1). $$CN \,.\, NT = CN \,.\, CT - CN^2 = AC^2 - CN^2$$ $$= AN \,.\, NA'.$$

COR. (2). Hence it follows that *tangents at the extremities of a common ordinate of an ellipse and its auxiliary circle meet the axis in the same point.*

For, if NP produced meet the auxiliary circle in Q, and the tangent at Q meet the axis in T',

$$CN \,.\, NT' = CQ^2 = AC^2,$$

therefore T' coincides with T.

And more generally it is evident that, *If any number of ellipses be described having the same major axis, and an ordinate be drawn cutting the ellipses, the tangents at the points of section will all meet the common axis in the same point.*

75. PROP. XV. *If the tangent at P meet the axis minor in t, and PN be the ordinate,*

$$Ct \,.\, PN = BC^2.$$

For, $$Ct : PN :: CT : NT \text{ (Fig. Art. 74)},$$

$$\therefore Ct \,.\, PN : PN^2 :: CT \,.\, CN : CN \,.\, NT$$
$$:: AC^2 : AN \,.\, NA' \text{ (Cor. 1, Art. 74)},$$
$$:: BC^2 : PN^2.$$
$$\therefore Ct \,.\, PN = BC^2.$$

76. PROP. XVI. *If the tangent and normal at P meet the axis major in T and G,*

$$CG \,.\, CT = SC^2.$$

The triangles CGg, CTt, in the figure of the next article, being similar,

$$CG : Cg :: Ct : CT,$$
$$\therefore CG \,.\, CT = Cg \,.\, Ct.$$

But, since t, S, g, S' are concyclic (Cor. Art. 68),

$$Cg \,.\, Ct = SC \,.\, CS' = SC^2;$$
$$\therefore CG \,.\, CT = SC^2.$$

COR. Since $CN \, . \, CT = AC^2$, and $PN \, . \, Ct = BC^2$,

$$CG : CN :: SC^2 : AC^2$$

and

$$Cg : PN :: SC^2 : BC^2.$$

We hence see that

$$NG : CN :: BC^2 : AC^2.$$

77. PROP. XVII. *If the normal at P meet the axes in G and g, and the diameter parallel to the tangent at P in F,*

$$PF \, . \, PG = BC^2, \text{ and } PF \, . \, Pg = AC^2.$$

Let PN, PM, perpendiculars on the axes, meet the diameter in K and L, and let the tangent at P meet the axes in T and t.

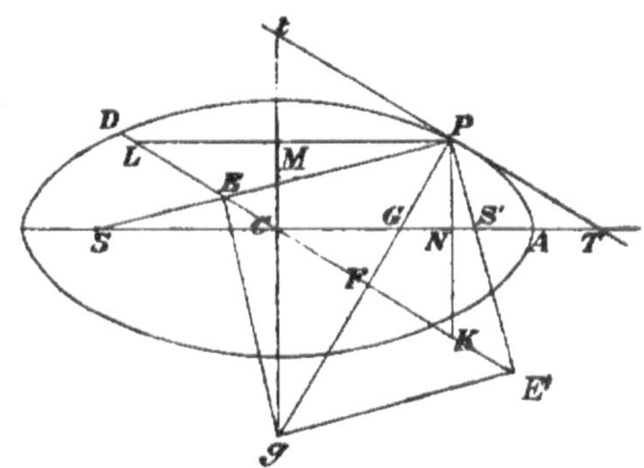

Then, since G, F, K, N are concyclic,

$$PF \, . \, PG = PN \, . \, PK = PN \, . \, Ct = BC^2.$$

Similarly, since L, M, F, g are concyclic,

$$PF \, . \, Pg = PM \, . \, PL = CN \, . \, CT = AC^2.$$

COR. If SP, $S'P$ meet the diameter DCD' parallel to the tangent at P in E and E',

$$PE = AC \text{ (Cor. 2, Art. 70)};$$

$$\therefore PF \, . \, Pg = PE^2 = PE'^2,$$

and hence it follows that the angles PEg, $PE'g$ are right angles.

78. PROP. XVIII. *If PCp be a diameter, QVQ' a chord parallel to the tangent at P and meeting Pp in V, and if the tangent at Q meet pP produced in T,*

$$CV \,.\, CT = CP^2.$$

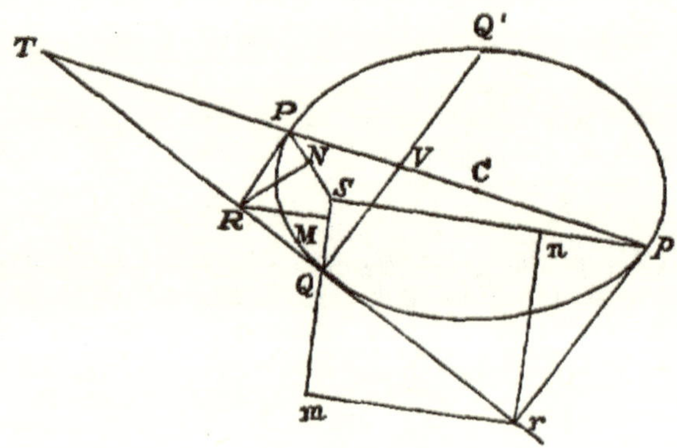

Let TQ meet the tangents at P and p in R and r, and S being a focus, join SP, SQ, Sp.

Let fall perpendiculars RN, RM, rn, rm upon these focal distances; then, since the angle $SPR = Spr$ (Cor. Art. 69),

$$RP : rp :: RN : rn$$
$$:: RM : rm \text{ (Cor. Art. 15),}$$
$$:: RQ : rQ;$$
$$:: PV : Vp.$$

Hence $$TP : Tp :: PV : Vp,$$

or $$CT - CP : CT + CP :: CP - CV : CP + CV;$$

$$\therefore CT : CP :: CP : CV,$$

or $$CT \,.\, CV = CP^2.$$

COR. 1. Hence, since CV and CP are the same for the point Q', the tangent at Q' passes through T.

COR. 2. Since $Tp : TP :: pV : VP$, it follows that $TPVp$ is harmonically divided.

It will be seen in a subsequent chapter that this is a particular case of a general theorem.

Properties of Conjugate Diameters.

79. PROP. XIX. *A diameter bisects all chords parallel to the tangents at its extremities.*

We have shewn in Art. 21, that, if QQ' be a chord of a conic, TQ, TQ' the tangents at Q, Q', and EPE' a tangent parallel to QQ', the length EE' is bisected at P.

Draw the diameter PCp; the tangent epe' at p is parallel to EPE' (Art. 69), and is therefore parallel to QQ'.

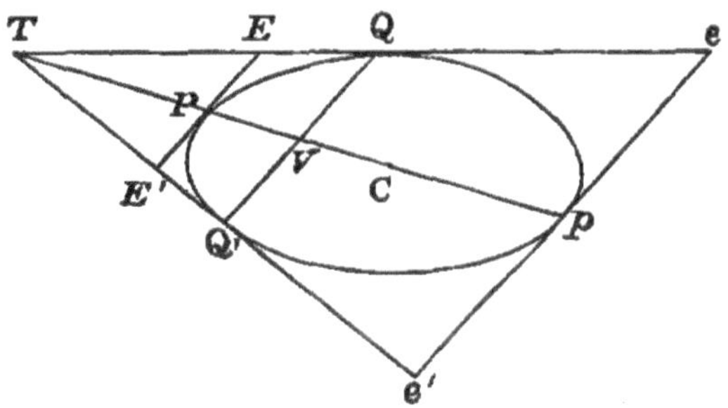

Hence $ep = pe'$, and P, p being the middle points of the parallels ee', EE' the line Pp passes through T, and moreover bisects QQ'.

Similarly, if any other chord qq' be drawn parallel to QQ' the tangents at q and q' will meet in pP produced, and qq' will be bisected by pP.

COR. Hence, if QQ', qq' be two chords parallel to the tangent at P, the chords Qq, $Q'q'$ will meet in CP or CP produced.

80. DEF. *The diameter DCd, drawn parallel to the tangent at P, is said to be conjugate to PCp.*

A diameter therefore bisects all chords parallel to its conjugate.

PROP. XX. *If the diameter DCd be conjugate to PCp, then will PCp be conjugate to DCd.*

Let the chord QVq be parallel to DCd, and therefore bisected by PC, and draw the diameter qCR.

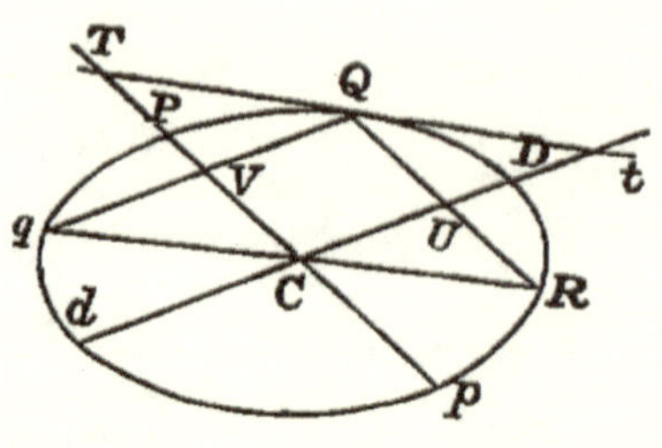

Join QR meeting CD in U; then $RC = Cq$, and $QV = Vq$;

$\therefore QR$ is parallel to CP.

Also $QU : UR :: qC : CR$, and therefore $QU = UR$.

That is, CD bisects the chords parallel to PCp; therefore PCp is conjugate to DCd.

DEF. *Chords drawn from the extremities of a diameter to any point of the ellipse are called supplemental chords.*

Thus qQ, RQ are supplemental chords, and hence it appears that supplemental chords are parallel to conjugate diameters.

DEF. *A line QV drawn from a point Q of an ellipse, parallel to the tangent at P and terminated by the diameter PCp, is called an ordinate of that diameter, and QVq is the double ordinate if QV produced meet the curve in q.*

81. *Any diameter is a mean proportional between the transverse axis and the focal chord parallel to the diameter.*

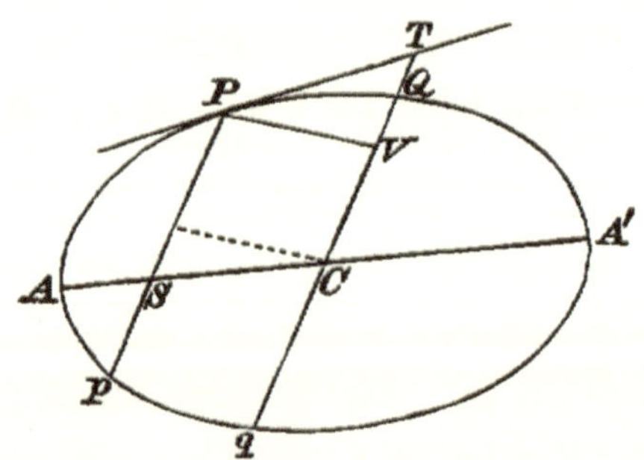

From Art. 70, it appears that if CQT parallel to SP meet in T the tangent at P,

$$CT = AC.$$

Draw PV parallel to the tangent at Q;

$$\text{then } CQ^2 = CV\,.\,CT = CV\,.\,AC;$$

but the diameter through C parallel to the tangent at Q bisects Pp (Art. 80),

$$\text{so that } Pp = 2CV;$$
$$\therefore Qq^2 = Pp \,.\, AA'.$$

82. PROP. XXI. *If PCp, DCd be conjugate diameters, and QV an ordinate of Pp,*

$$QV^2 : PV \,.\, Vp :: CD^2 : CP^2.$$

Let the tangent at Q (Fig. Art. 80) meet CP, CD produced in T and t, and draw QU parallel to CP and meeting CD in U.

Then $$CP^2 = CV \,.\, CT,$$
and $$CD^2 = CU \,.\, Ct = QV \,.\, Ct;$$
$$\therefore CD^2 : CP^2 :: QV \,.\, Ct : CV \,.\, CT$$
$$:: QV^2 : CV \,.\, VT,$$
and $$CV \,.\, VT = CV \,.\, CT - CV^2 = CP^2 - CV^2$$
$$= PV \,.\, Vp,$$
$$\therefore CD^2 : CP^2 :: QV^2 : PV \,.\, Vp.$$

83. PROP. XXII. *If ACA', BCB' be a pair of conjugate diameters, PCP', DCD' another pair, and if PN, DM be ordinates of ACA',*

$$CN^2 = AM \,.\, MA', \quad CM^2 = AN \,.\, NA',$$
$$CM : PN :: AC : BC,$$
and $$DM : CN :: BC : AC.$$

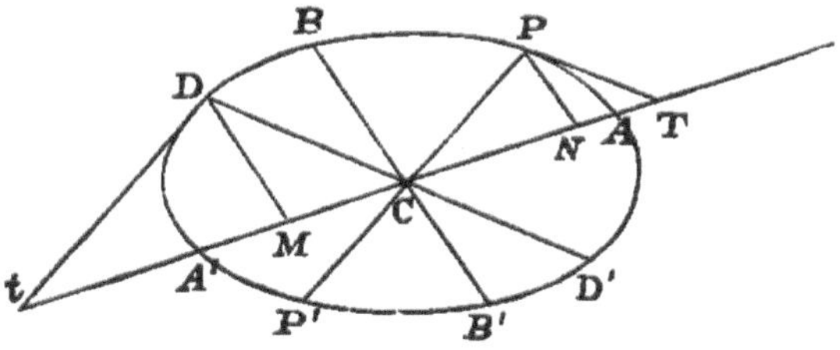

Let the tangents at P and D meet ACA' in T and t.

Then $$CN \,.\, CT = AC^2 = CM \,.\, Ct;$$

hence
$$\begin{aligned} CM : CN &:: CT : Ct \\ &:: PT : CD \\ &:: PN : DM \\ &:: CN : Mt, \end{aligned}$$

$$\therefore CN^2 = CM \,.\, Mt = AC^2 - CM^2 = AM \,.\, MA',$$

and similarly, $$CM^2 = AN \,.\, NA'.$$

Also $$DM^2 : AM \,.\, MA' :: BC^2 : AC^2,$$

$$\therefore DM : CN :: BC : AC,$$

and similarly $$CM : PN :: AC : BC.$$

COR. We have shewn in the course of the proof that

$$CN^2 + CM^2 = AC^2.$$

By similar reasoning it appears that if Pn, Dm, be ordinates of BCB',

$$Cn^2 + Cm^2 = BC^2;$$

$$\therefore PN^2 + DM^2 = BC^2.$$

It should be noticed that these relations are shewn to be true when ACA', BCB' are any conjugate diameters, including of course the principal axes.

84. PROP. XXIII. If CP, CD be conjugate semi-diameters, and AC, BC the principal semi-diameters,

$$CP^2 + CD^2 = AC^2 + BC^2.$$

From the preceding article,

$$CN^2 + CM^2 = AC^2,$$

and $$PN^2 + DM^2 = BC^2;$$

also ACB being in this case a right angle,

$$PN^2 + CN^2 = CP^2,$$

and $$DM^2 + CM^2 = CD^2,$$

$$\therefore CP^2 + CD^2 = AC^2 + BC^2.$$

85. DEF. *If the ordinate NP of a point, when produced, meets the auxiliary circle in Q, the angle ACQ is called the eccentric angle of the point P.*

Prop. XXIV. *If CP, CD be conjugate semi-diameters, the difference between the eccentric angles of P and D is a right angle.*

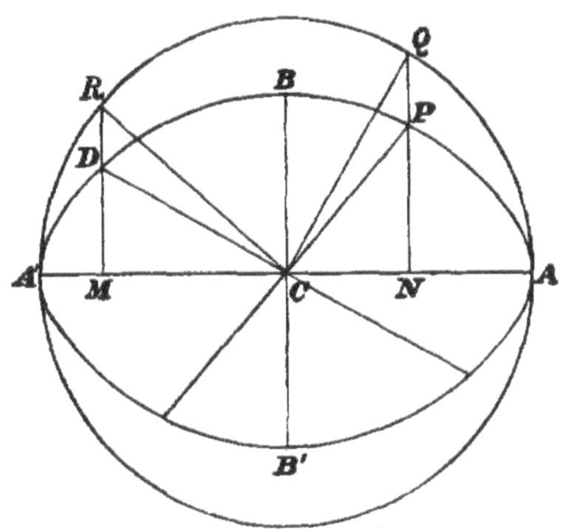

From Art. 67, $RM : DM :: AC : BC$

and, from Art. 83, $CN : DM :: AC : BC$

$$\therefore RM = CN, \text{ and similarly, } QN = CM.$$

$\therefore$ The triangles QCN, CRM are equal, and the angles QCN, RCM are complementary.

$$\therefore QCR \text{ is a right angle.}$$

86. Prop. XXV. *If the normal at P meet the principal axes in G and g,*

$$PG : CD :: BC : AC,$$

and $$Pg : CD :: AC : BC.$$

For, the triangles DCM, PGN being similar,

$$PG : CD :: PN : CM$$
$$:: BC : AC.$$

So also Pgn and DCM are similar, and

$$Pg : CD :: Pn : DM$$
$$:: AC : BC.$$

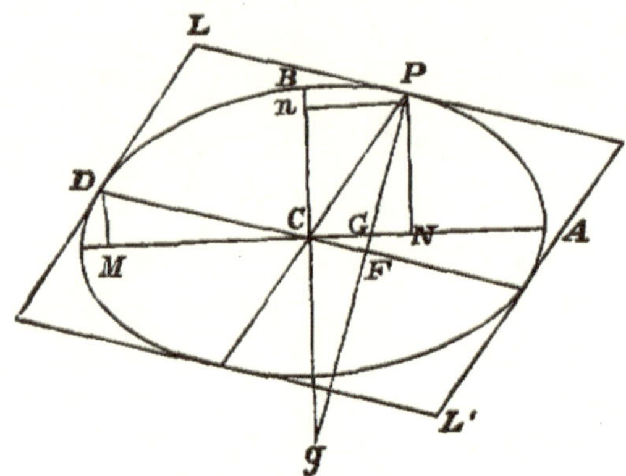

Hence it follows that

$$PG \,.\, Pg = CD^2.$$

87. Prop. XXVI. *The parallelogram formed by the tangents at the ends of conjugate diameters is equal to the rectangle contained by the principal axes.*

For, taking the preceding figure,

$$PG : BC :: CD : AC;$$

but $$PG : BC :: BC : PF \text{ (Art. 77)},$$

$$\therefore CD : AC :: BC : PF,$$

and $$CD \,.\, PF = AC \,.\, BC,$$

whence the theorem stated.

88. Prop. XXVII. *If SP, $S'P$ be the focal distances of P, and CD be conjugate to CP,*

$$SP \,.\, S'P = CD^2,$$

and $$SY : SP :: BC : CD.$$

Let CD meet SP, $S'P$ in E and E', and the normal at P in F; then SPY, PEF, and $S'PY'$ are similar triangles;

$$\therefore SP : SY :: PE : PF,$$

and $$S'P : S'Y' :: PE : PF;$$

$$\therefore SP \,.\, S'P : SY \,.\, S'Y' :: PE^2 : PF^2$$
$$:: AC^2 : PF^2$$
$$:: CD^2 : BC^2 \text{ (Art. 87)};$$
$$\therefore SP \,.\, S'P = CD^2.$$

Also $$SY : SP :: PF : PE :: PF : AC,$$
$$\therefore SY : SP :: BC : CD.$$

89. PROP. XXVIII. *If the tangent at P meet a pair of conjugate diameters in T and T', and CD be conjugate to CP,*

$$PT \,.\, PT' = CD^2.$$

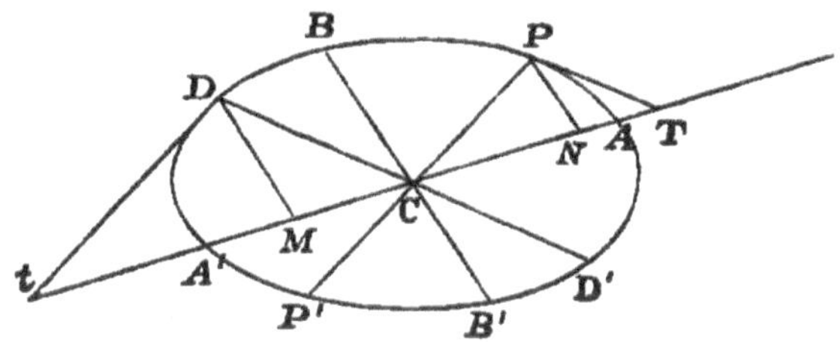

From the figure

$$PT : PN :: CD : DM;$$

and, if TP produced meet CB in T',

$$PT' : CN :: CD : CM;$$
$$\therefore PT \,.\, PT' : PN \,.\, CN :: CD^2 : DM \,.\, CM.$$

But $$PN \,.\, CN = DM \,.\, CM \text{ (Art. 83)},$$
$$\therefore PT \,.\, PT' = CD^2.$$

COR. Let TQU be the tangent at the other end of the chord PNQ, meeting CB' produced in U; and let CE be the semi-diameter parallel to TQ.

Then $$TP : TQ :: PT' : QU,$$
$$\therefore TP^2 : TQ^2 :: PT \,.\, PT' : QT \,.\, QU$$
$$:: CD^2 : CE^2,$$

that is, *the two tangents drawn from any point are in the ratio of the parallel diameters.*

In a similar manner it can be shewn that, if the tangent at P meet the tangents at the ends of a diameter ACA' in T and T',

$$PT \,.\, PT' = CD^2,$$

CD being conjugate to CP,

and
$$AT \,.\, A'T' = CB^2,$$

CB being conjugate to ACA'.

90. *Equi-conjugate diameters.*

PROP. XXIX. *The diagonals of the rectangle formed by the principal axes are equal and conjugate diameters.*

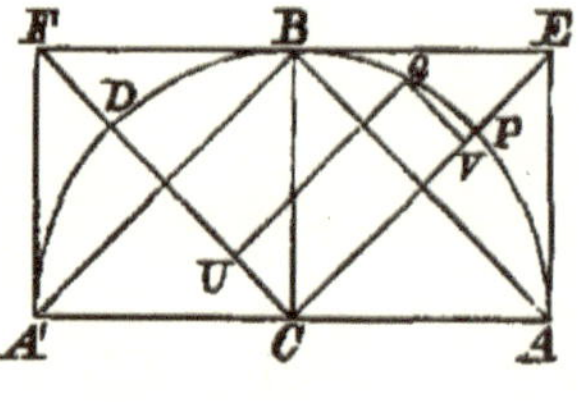

For, joining AB, $A'B$, these lines are parallel to the diagonals CF, CE; and, AB, $A'B$ being supplemental chords, it follows that CD, CP are conjugate to each other. Moreover, they are equally inclined to the axes, and are therefore of equal length.

COR. 1. If QV, QU be drawn parallel to the equi-conjugate diameters, meeting them in V and U,

$$QV^2 : CP^2 - CV^2 :: CD^2 : CP^2;$$

$$\therefore\ QV^2 = CP^2 - CV^2 = PV \,.\, VP',$$

if P' be the other end of the diameter PCP'.

Hence
$$QV^2 + QU^2 = CP^2.$$

COR. 2.
$$CP^2 + CD^2 = AC^2 + BC^2 \text{ (Art. 84)};$$

$$\therefore\ 2CP^2 = AC^2 + BC^2.$$

91. PROP. XXX. *Pairs of tangents at right angles to each other intersect on a fixed circle.*

The two tangents being TP, TP', let $S'P$ produced meet SY the perpendicular on TP in K.

Then the angle $PTK = STP = S'TP'$;

$$\therefore\ S'TK \text{ is a right angle.}$$

Hence
$$4AC^2 = S'K^2 = S'T^2 + TK^2$$

$$= S'T^2 + ST^2$$

$$= 2CT^2 + 2CS^2 \text{ (Euclid, II. 12 and 13)};$$

$$\therefore\ CT^2 = AC^2 + BC^2,$$

and T lies on a fixed circle, of which C is the centre.

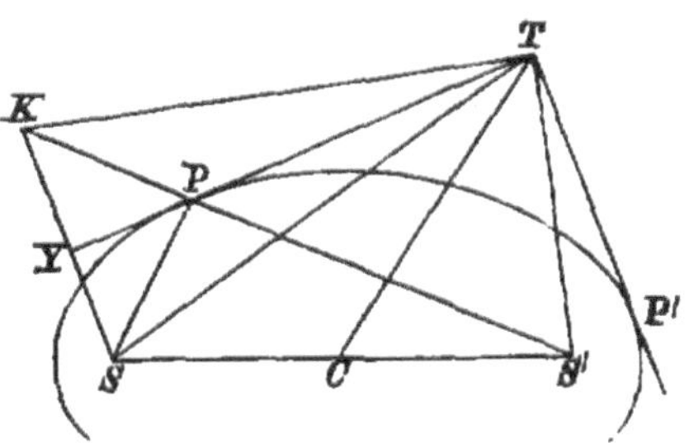

This circle is called the *Director Circle* of the Ellipse, and it will be seen that when the ellipse, by the elongation of SC from S is transformed into a parabola, the director circle merges into the directrix of the parabola.

COR. If XQ is the tangent to the director circle from the foot of the directrix,

$$\begin{aligned} XQ^2 &= CX^2 - CQ^2 = CX^2 - CA^2 - CB^2 \\ &= CX^2 - SC \,.\, CX - SC \,.\, SX \text{ (Arts. 61 and 63)}, \\ &= CX \,.\, SX - SC \,.\, SX = SX^2. \end{aligned}$$

$$\therefore\ XQ = SX,$$

and hence it follows that *the directrix is the radical axis of the director circle and of a point circle at the focus.*

92. PROP. XXXI. *The rectangles contained by the segments of any two chords which intersect each other are in the ratio of the squares of the parallel diameters.*

Through any point O in a chord OQQ' draw the diameter ORR', and let CD be parallel to QQ', and CP conjugate to CD, bisecting QQ' in V.

Draw RU parallel to CD.

Then $$CD^2 - RU^2 : CU^2 :: CD^2 : CP^2 \text{ (Art. 82)},$$
$$:: CD^2 - QV^2 : CV^2.$$

But $$RU^2 : CU^2 :: OV^2 : CV^2;$$
$$\therefore\ CD^2 : CU^2 :: CD^2 + OV^2 - QV^2 : CV^2$$

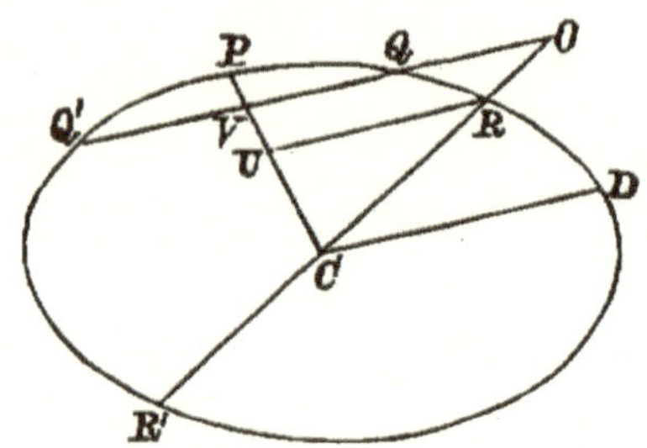

$$\text{or} \qquad CD^2 : CD^2 + OV^2 - QV^2 :: CU^2 : CV^2$$
$$:: CR^2 : CO^2;$$
$$\therefore\ CD^2 : OV^2 - QV^2 :: CR^2 : CO^2 - CR^2,$$
$$\text{or} \qquad CD^2 : OQ \,.\, OQ' :: CR^2 : OR \,.\, OR'.$$

Similarly, if Oqq' be any other chord through O, and Cd the parallel semi-diameter,

$$Cd^2 : Oq \,.\, Oq' :: CR^2 : OR \,.\, OR';$$
$$\therefore\ OQ \,.\, OQ' : Oq \,.\, Oq' :: CD^2 : Cd^2.$$

This may otherwise be expressed thus,

The ratio of the rectangles of the segments depends only on the directions in which they are drawn.

The proof is the same if the point O be within the ellipse.

93. PROP. XXXII. *If a circle intersect an ellipse in four points, the several pairs of the chords of intersection are equally inclined to the axes.*

For if QQ', qq' be a pair of the chords of intersection, and if these meet in O, or be produced to meet in O, the rectangles $OQ \,.\, OQ'$, $Oq \,.\, Oq'$ are proportional to the squares on the parallel diameters.

But these rectangles are equal since QQ', qq' are chords of a circle.

Therefore the parallel diameters are equal, and, since equal diameters are equally inclined to the axes, it follows that the chords QQ', qq' are equally inclined to the axes.

Conversely, if two chords, not parallel, be equally inclined to the axes a circle can be drawn through their extremities.

For, as in Art. 92, if OQQ', Oqq' be two chords, and CD, Cd the parallel semi-diameters,

$$OQ \,.\, OQ' : Oq \,.\, Oq' :: CD^2 : Cd^2;$$

but, if CD and Cd be equally inclined to the axes, they are equal, and

$$\therefore OQ \,.\, OQ' = Oq \,.\, Oq',$$

and the points Q, Q', q, q' are concyclic.

EXAMPLES.

1. If the tangent at B meet the latus rectum produced in D, CDX is a right angle.

2. If PCp be a diameter, and the focal distance pS produced meet the tangent at P in T, $SP = ST$.

3. If the normal at P meet the axis minor in G' and $G'N$ be the perpendicular from G' on SP, then $PN = AC$.

4. The tangent at P bisects any straight line perpendicular to AA' and terminated by AP, $A'P$, produced if necessary.

5. Draw a tangent to an ellipse parallel to a given line.

6. SR being the semi-latus rectum, if RA meet the directrix in E, and $S'E$ meet the tangent at A in T,

$$AT = AS.$$

7. Prove that $SY : SP :: SR : PG$.

Find where the angle SPS' is greatest.

8. If two points E and E' be taken in the normal PG such that $PE = PE' = CD$, the loci of E and E' are circles.

9. If from the focus S' a line be drawn parallel to SP, it will meet the perpendicular SY in the circumference of a circle.

10. If the normal at P meet the axis major in G, prove that PG is an harmonic mean between the perpendiculars from the foci on the tangent at P.

11. The straight line NQ is drawn parallel to AP to meet CP in Q; prove that AQ is parallel to the tangent at P.

12. The locus of the intersection with the ordinate of the perpendicular from the centre on the tangent is an ellipse.

13. If a rectangle circumscribes an ellipse, its diagonals are the directions of conjugate diameters.

14. If tangents TP, TQ be drawn at the extremities, P, Q of any focal chord of an ellipse, prove that the angle PTQ is half the supplement of the angle which PQ subtends at the other focus.

15. If Y, Z be the feet of the perpendiculars from the foci on the tangent at P; prove that Y, N, Z, C are concyclic.

16. If AQ be drawn from one of the vertices perpendicular to the tangent at any point P, prove that the locus of the point of intersection of PS and QA produced will be a circle.

17. The straight lines joining each focus to the foot of the perpendicular from the other focus on the tangent at any point meet on the normal at the point and bisect it.

18. If two circles touch each other internally, the locus of the centres of circles touching both is an ellipse whose foci are the centres of the given circles.

19. The subnormal at any point P is a third proportional to the intercept of the tangent at P on the major axis and half the minor axis.

20. If the normal at P meet the axis major in G and the axis minor in g, $Gg : Sg :: SA : AX$, and if the tangent meet the axis minor in t,

$$St : tg :: BC : CD.$$

21. If the normal at a point P meet the axis in G, and the tangent at P meet the axis in T, prove that

$$TQ : TP :: BC : PG,$$

Q being the point where the ordinate at P meets the auxiliary circle.

22. If the tangent at any point P meet the tangent at the extremities of the axis AA' in F and F', prove that the rectangle AF, $A'F'$ is equal to the square on the semi-axis minor.

23. TP, TQ are tangents; prove that a circle can be described with T as centre so as to touch SP, HP, SQ, and HQ, or these lines produced, S and H being the foci.

24. If two equal and similar ellipses have the same centre, their points of intersection are at the extremities of diameters at right angles to one another.

25. The external angle between any two tangents to an ellipse is equal to the semi-sum of the angles which the chord joining the points of contact subtends at the foci.

26. The tangent at any point P meets the axes in T and t; if S be a focus the angles PSt, STP are equal.

27. A conic is drawn touching an ellipse at the extremities A, B of the axes, and passing through the centre C of the ellipse; prove that the tangent at C is parallel to AB.

28. The tangent at any point P is cut by any two conjugate diameters in T, t, and the points T, t are joined with the foci S, H respectively; prove that the triangles SPT, HPt are similar to each other.

29. If the diameter conjugate to CP meet SP, and HP (or these produced) in E and E', prove that SE is equal to HE', and that the circles which circumscribe the triangles SCE, HCE', are equal to one another.

30. PG is a normal, terminating in the major axis; the circle, of which PG is a diameter, cuts SP, HP, in K, L, respectively: prove that KL is bisected by PG, and is perpendicular to it.

31. Tangents are drawn from any point in a circle through the foci, prove that the lines bisecting the angles between the several pairs of tangents all pass through a fixed point.

32. If a quadrilateral circumscribe an ellipse, the angles subtended by opposite sides at one of the foci are together equal to two right angles.

33. If the normal at P meet the axis minor in G, and if the tangent at P meet the tangent at the vertex A in V, shew that

$$SG : SC :: PV : VA.$$

34. P, Q are points in two confocal ellipses, at which the line joining the common foci subtends equal angles; prove that the tangents at P, Q are inclined at an angle which is equal to the angle subtended by PQ at either focus.

35. The transverse axis is the greatest and the conjugate axis the least of all the diameters.

36. Prove that the locus of the centre of the circle inscribed in the triangle SPS' is an ellipse.

37. If the tangent and ordinate at P meet the transverse axis in T and N, prove that any circle passing through N and T will cut the auxiliary circle orthogonally.

38. If SY, $S'Y'$ be the perpendiculars from the foci on the tangent at a point P, and PN the ordinate, prove that

$$PY : PY' :: NY : NY'.$$

39. If a circle, passing through Y and Z, touch the major axis in Q, and that diameter of the circle, which passes through Q, meet the tangent in P, then $PQ = BC$.

40. From the centre of two concentric circles a straight line is drawn to cut them in P and Q; from P and Q straight lines are drawn parallel to two given lines at right angles. Shew that the locus of their point of intersection is an ellipse.

41. From any two points P, Q on an ellipse four lines are drawn to the foci S, S': prove that $SP \,.\, S'Q$ and $SQ \,.\, S'P$ are to one another as the squares of the perpendiculars from a focus on the tangents at P and Q.

42. Two conjugate diameters are cut by the tangent at any point P in M, N; prove that the area of the triangle CPM varies inversely as that of the triangle CPN.

43. If P be any point on the curve, and AV be drawn parallel to PC to meet the conjugate CD in V, prove that the areas of the triangles CAV, CPN are equal, PN being the ordinate.

44. Two tangents to an ellipse intersect at right angles; prove that the sum of the squares on the chords intercepted on them by the auxiliary circle is constant.

45. Prove that the distance between the two points on the circumference, at which a given chord, not passing through the centre, subtends the greatest and least angles, is equal to the diameter which bisects that chord.

46. The tangent at P intersects a fixed tangent in T; if S is the focus and a line be drawn through S perpendicular to ST, meeting the tangent at P in Q, shew that the locus of Q is a straight line touching the ellipse.

47. Shew that, if the distance between the foci be greater than the length of the axis minor, there will be four positions of the tangent, for which the area of the triangle, included between it and the straight lines drawn from the centre of the curve to the feet of the perpendiculars from the foci on the tangent, will be the greatest possible.

48. Two ellipses whose axes are equal, each to each, are placed in the same plane with their centres coincident, and axes inclined to each other. Draw their common tangents.

49. An ellipse is inscribed in a triangle, having one focus at the orthocentre; prove that the centre of the ellipse is the centre of the nine-point circle of the triangle and that its transverse axis is equal to the radius of that circle.

50. The tangent at any point P of a circle meets the tangent at a fixed point A in T, and T is joined with B the extremity of the diameter passing through A; the locus of the point of intersection of AP, BT is an ellipse.

51. The ordinate NP at a point P meets, when produced, the circle on the major axis in Q. If S be a focus of the ellipse, prove that $SQ : SP ::$ the axis major : the chord of the circle through Q and S, and that the diameter of the ellipse parallel to SP is equal to the same chord.

52. If the perpendicular from the centre C on the tangent at P meet the focal distance SP produced in R, the locus of R is a circle, the diameter of which is equal to the axis major.

53. A perfectly elastic billiard ball lies on an elliptical billiard table, and is projected in any direction along the table: shew that all the lines in which it moves after each successive impact touch an ellipse or an hyperbola confocal with the billiard table.

54. Shew that a circle can be drawn through the foci and the intersections of any tangent with the tangents at the vertices.

55. If CP, CD be conjugate semi-diameters, and a rectangle be described so as to have PD for a diagonal and its sides parallel to the axes, the other angular points will be situated on two fixed straight lines passing through the centre C.

56. If the tangent at P meet the minor axis in T, prove that the areas of the triangles SPS', STS' are in the ratio of the squares on CD and ST.

57. Find the locus of the centre of the circle touching the transverse axis, SP, and $S'P$ produced.

58. In an ellipse SQ and $S'Q$, drawn perpendicularly to a pair of conjugate diameters, intersect in Q; prove that the locus of Q is a concentric ellipse.

59. If the ordinate NP meet the auxiliary circle in Q, the perpendicular from S on the tangent at Q is equal to SP.

60. If PT, QT be tangents at corresponding points of an ellipse and its auxiliary circle, shew that

$$PT : QT :: BC : PF.$$

61. If CQ be conjugate to the normal at P, then is CP conjugate to the normal at Q.

62. PQ is one side of a parallelogram described about an ellipse, having its sides parallel to conjugate diameters, and the lines joining P, Q to the foci intersect in D, E; prove that the points D, E and the foci are concyclic.

63. If the centre, a tangent, and the transverse axis be given, prove that the directrices pass each through a fixed point.

64. The straight line joining the feet of perpendiculars from the focus on two tangents is at right angles to the line joining the intersection of the tangents with the other focus.

65. A circle passes through a focus, has its centre on the major axis of the ellipse, and touches the ellipse: shew that the straight line from the focus to the point of contact is equal to the latus rectum.

66. Prove that the perimeter of the quadrilateral formed by the tangent, the perpendiculars from the foci, and the transverse axis, will be the greatest possible when the focal distances of the point of contact are at right angles to each other.

67. Given a focus, the length of the transverse axis, and that the second focus lies on a straight line, prove that the ellipse will touch two fixed parabolas having the given focus for focus.

68. Tangents are drawn from a point on one of the equi-conjugate diameters; prove that the point, the centre, and the two points of contact are concyclic.

69. If PN be the ordinate of P, and if with centre C and radius equal to PN a circle be described intersecting PN in Q, prove that the locus of Q is an ellipse.

70. If AQO be drawn parallel to CP, meeting the curve in Q and the minor axis in O, $2CP^2 = AO \,.\, AQ$.

71. PS is a focal distance; CR is a radius of the auxiliary circle parallel to PS, and drawn in the direction from P to S; SQ is a perpendicular on CR: shew that the rectangle contained by SP and QR is equal to the square on half the minor axis.

72. If a focus be joined with the point where the tangent at the nearer vertex intersects any other tangent, and perpendiculars be let fall from the other focus on the joining line and on the last-mentioned tangent, prove that the distance between the feet of these perpendiculars is equal to the distance from either focus to the remoter vertex.

73. A parallelogram is described about an ellipse; if two of its angular points lie on the directrices, the other two will lie on the auxiliary circle.

74. From a point in the auxiliary circle straight lines are drawn touching the ellipse in P and P'; prove that SP is parallel to $S'P'$.

75. Find the locus of the points of contact of tangents to a series of confocal ellipses from a fixed point in the axis major.

76. A series of confocal ellipses intersect a given straight line; prove that the locus of the points of intersection of the pairs of tangents drawn at the extremities of the chords of intersection is a straight line at right angles to the given straight line.

77. Given a focus and the length of the major axis; describe an ellipse touching a given straight line and passing through a given point.

78. Given a focus and the length of the major axis; describe an ellipse touching two given straight lines.

79. Find the positions of the foci and directrices of an ellipse which touches at two given points P, Q, two given straight lines PO, QO, and has one focus on the line PQ, the angle POQ being less than a right angle.

80. Through any point P of an ellipse are drawn straight lines APQ, $A'PR$, meeting the auxiliary circle in Q, R, and ordinates Qq, Rr are drawn to the transverse axis; prove that, L being an extremity of the latus rectum,

$$Aq \,.\, A'r : Ar \,.\, A'q :: AC^2 : SL^2.$$

81. If a tangent at a point P meet the major axis in T, and the perpendiculars from the focus and centre in Y and Z, then

$$TY^2 : PY^2 :: TZ : PZ.$$

82. An ellipse slides between two lines at right angles to each other; find the locus of its centre.

83. TP, TQ are two tangents, and CP', CQ' are the radii from the centre respectively parallel to these tangents, prove that $P'Q'$ is parallel to PQ.

84. The tangent at P meets the minor axis in t; prove that

$$St \,.\, PN = BC \,.\, CD.$$

85. If the circle, centre t, and radius tS, meet the ellipse in Q, and QM be the ordinate, prove that

$$QM : PN :: BC : BC + CD.$$

86. Perpendiculars SY, $S'Y'$ are let fall from the foci upon a pair of tangents TY, TY'; prove that the angles STY, $S'TY'$ are equal to the angles at the base of the triangle YCY'.

87. PQ is the chord of an ellipse normal at P, LCL' the diameter bisecting it, shew that PQ bisects the angle LPL' and that $LP + PL'$ is constant.

88. ABC is an isosceles triangle of which the side AB is equal to the side AC. BD, BE drawn on opposite sides of BC and equally inclined to it meet AC in D and E. If an ellipse is described round BDE having its axis minor parallel to BC, then AB will be a tangent to the ellipse.

89. If A be the extremity of the major axis and P any point on the curve, the bisectors of the angles PSA, $PS'A$ meet on the tangent at P.

90. If two ellipses intersect in four points, the diameters parallel to a pair of the chords of intersection are in the same ratio to each other.

91. From any point P of an ellipse a straight line PQ is drawn perpendicular to the focal distance SP, and meeting in Q the diameter conjugate to that through P; shew that PQ varies inversely as the ordinate of P.

92. If a tangent to an ellipse intersect at right angles a tangent to a confocal ellipse, the point of intersection lies on a fixed circle.

93. If from a point T in the director circle of an ellipse tangents TP, TP' are drawn, the line joining T with the intersection of the normals at P and P' passes through C.

94. Through the middle point of a focal chord a straight line is drawn at right angles to it to meet the axis in R; prove that SR bears to SC the duplicate ratio of the chord to the diameter parallel to it, S being the focus and C the centre.

95. The tangent at a point P meets the auxiliary circle in Q' to which corresponds Q on the ellipse; prove that the tangent at Q cuts the auxiliary circle in the point corresponding to P.

96. If a chord be drawn to a series of concentric, similar, and similarly situated ellipses, and meet one in P and Q, and if on PQ as diameter a circle be described meeting that ellipse again in RS, shew that RS is constant in position for all the ellipses.

97. An ellipse touches the sides of a triangle; prove that if one of its foci move along the arc of a circle passing through two of the angular points of the triangle, the other will move along the arc of a circle through the same two angular points.

98. The normal at a point P of an ellipse meets the conjugate axis in K, and a circle is described with centre K and passing through the foci S and H. The lines SQ, HQ, drawn through any point Q of this circle, meet the tangent at P in T and t; prove that T and t lie on a pair of conjugate diameters.

99. If SP, $S'Q$ be parallel focal distances drawn towards the same parts, the tangents at P and Q intersect on the auxiliary circle.

100. Having given one focus, one tangent and the eccentricity of an ellipse, prove that the locus of the other focus is a circle.

101. PSQ is a focal chord of an ellipse, and pq is any parallel chord; if PQ meet in T the tangent at p,

$$pq : PQ :: Sp : ST.$$

102. If an ellipse be inscribed in a quadrilateral so that one focus is equidistant from the four vertices, the other focus must be at the intersection of the diagonals.

103. If a pair of conjugate diameters of an ellipse be produced to meet either directrix, prove that the orthocentre of the triangle so formed is the corresponding focus of the curve.

104. A pair of conjugate diameters intercept, on the tangent at either vertex, a length which subtends supplementary angles at the foci.

105. The straight lines TP, TQ are the tangents at the points P, Q of an ellipse; one circle touches TP at P and meets TQ in Q and Q', and another circle touches TQ at Q and meets TP in P and P'; prove that PQ' and $P'Q$ are parallel, and that they are divided in the same ratio by the ellipse.

106. If the normals at P and D meet in E, prove that EC is perpendicular to PD, and that the straight line joining C to the centroid of the triangle EPD bisects the line joining E to T, the point of intersection of the tangents at P and D.

107. A chord PQ, normal at P, meets the directrices in K and L, and the tangents at P and Q meet in T; prove that PK and QL subtend equal angles at T, and that KL subtends at T an angle which is half the sum of the angles subtended by SS' at the ends of the chord.

108. The tangent at the point P meets the directrices in E and F; prove that the other tangents from E and F intersect on the normal at P.

109. If the tangent at any point meets a pair of conjugate diameters in T and T', prove that TT' subtends supplementary angles at the foci.

110. PSQ, $PS'R$ are focal chords; prove that the tangent at P and the chord QR cut the major axis at equal distances from the centre.

CHAPTER IV.

THE HYPERBOLA.

DEFINITION.

An hyperbola is the curve traced by a point which moves in such a manner, that its distance from a given point is in a constant ratio of greater inequality to its distance from a given straight line.

Tracing the Curve.

94. Let S be the focus, EX the directrix, and A the vertex.

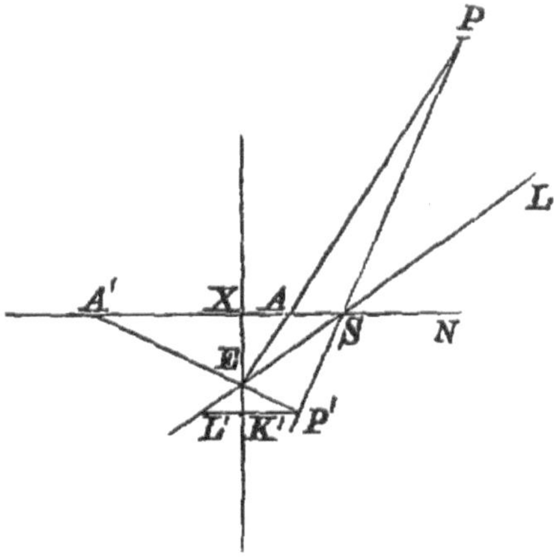

Then, as in Art. 1, any number of points on the curve may be obtained by taking successive positions of E on the directrix.

In SX produced, find a point A' such that

$$SA' : A'X :: SA : AX,$$

then A' is the other vertex as in the ellipse, and, the eccentricity being greater than unity, the points A and A' are evidently on opposite sides of the directrix.

Find the point P corresponding to E, and let $A'E$, PS produced meet in P', then, if $P'K'$ perpendicular to the directrix meet SE produced in L',

$$P'L' : P'K' :: SA' : A'X :: SA : AX,$$

and the angle

$$P'L'S = L'SX = L'SP';$$

$$\therefore SP' = P'L'.$$

Hence P' is a point in the curve, and PSP' is a focal chord.

Following out the construction we observe that, since SA is greater than AX, there are two points on the directrix, e and e', such that Ae and Ae' are each equal to AS.

If E coincide with e, the angle

$$QSL = LSN = ASe = AeS.$$

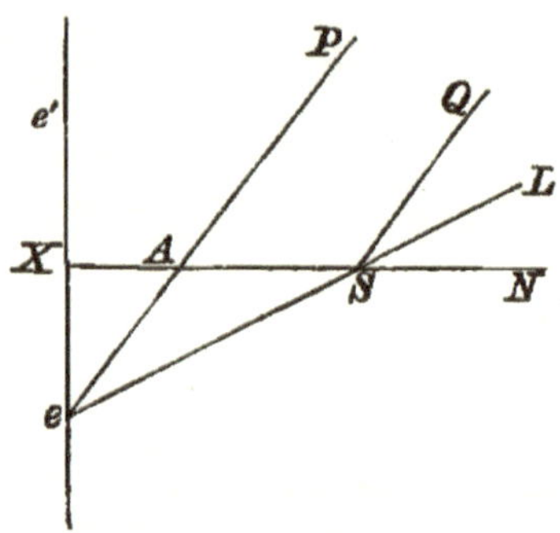

Hence SQ, AP are parallel, and the corresponding point of the curve is at an infinite distance; and similarly the curve tends to infinity in the direction Ae'.

Further, the angle ASE is less or greater than AES, according as the point E is, or is not, between e and e'.

Hence, when E is below e, the curve lies above the axis, to the right of the directrix; when between e and X, below the axis to the left; when between

X and e', above the axis to the left; and when above e', below the axis to the right. Hence a general idea can be obtained of the form of the curve, tending to infinity in four directions, as in the figure of Art. 102.

DEFINITIONS.

The line AA' is called the transverse axis of the hyperbola.

The middle point, C, of AA' is the centre.

Any straight line, drawn through C and terminated by the curve, is called a diameter.

95. PROP. I. *If P be any point of an hyperbola, and AA' its transverse axis, and if $A'P$, and PA produced, (or PA and PA' produced) meet the directrix in E and F, EF subtends a right angle at the focus.*

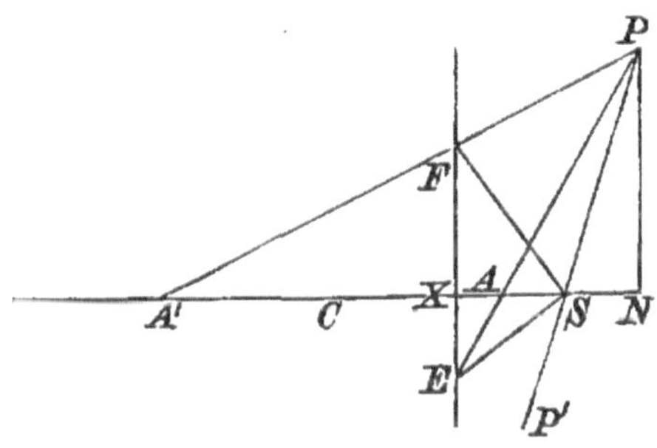

By the theorem of Art. 4, ES' bisects the angle ASP' and FS bisects ASP;

$$\therefore ESF \text{ is a right angle.}$$

SAA' being a focal chord, this is a particular case of the theorem of Art. 6.

96. PROP. II. *If PN be the ordinate of a point P, and ACA' the transverse axis, PN^2 is to $AN \,.\, NA'$ in a constant ratio.*

Join AP, $A'P$, meeting the directrix in E and F.

Then $$PN : AN :: EX : AX,$$

and $$PN : A'N :: FX : A'X;$$

$$\therefore PN^2 : AN \,.\, NA' :: EX \,.\, FX : AX \,.\, A'X$$
$$:: SX^2 : AX \,.\, A'X,$$

since ESF is a right angle; that is, PN^2 is to $AN \,.\, NA'$, in a constant ratio.

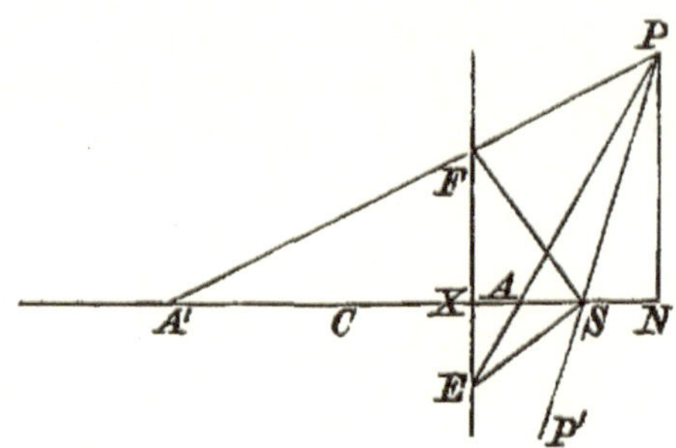

Through C, the middle point of AA', draw CB at right angles to the axis, and such that

$$BC^2 : AC^2 :: SX^2 : AX \,.\, A'X;$$

then
$$PN^2 : AN \,.\, NA' :: BC^2 : AC^2$$

or
$$PN^2 : CN^2 - AC^2 :: BC^2 - AC^2$$

Cor. If PM be the perpendicular from P to BC

$$PM = CN, \text{ and } PN = CM;$$

$$\therefore CM^2 : PM^2 - AC^2 :: BC^2 : AC^2$$

or
$$CM^2 : BC^2 :: PM^2 - AC^2 : AC^2$$

$$\therefore CM^2 + BC^2 : BC^2 :: PM^2 : AC^2$$

or
$$PM^2 : CM^2 + BC^2 :: AC^2 : BC^2$$

97. If we describe the circle on AA' as diameter, which we may term, for convenience, *the auxiliary circle*, the rectangle $AN \,.\, NA'$ is equal to the square on the tangent to the circle from N.

Hence the preceding theorem may be thus expressed:

The ordinate of an hyperbola is to the tangent from its foot to the auxiliary circle in the ratio of the conjugate to the transverse axis.

Def. *If CB' be taken equal to CB, on the other side of the axis, the line BCB' is called the conjugate axis.*

The two lines AA', BB' are the principal axes of the curve.

When these lines are equal, the hyperbola is said to be equilateral, or rectangular.

The lines AA', BB' are sometimes called major and minor axes, but, as AA' is not necessarily greater than BB', these terms cannot with propriety be generally employed.

If a point N' be taken on CA' produced, such that $CN' = CN$, the corresponding ordinate $P'N' = PN$, and therefore it follows that the curve is symmetrical with regard to BCB', and that there is another focus and directrix, corresponding to the vertex A'.

98. Prop. III. *If ACA' be the transverse axis, C the centre, S one of the foci, and X the foot of the directrix,*

$$CS : CA :: CA : CX :: SA : AX,$$

and

$$CS : CX :: CS^2 : CA^2.$$

Interchanging the positions of S and X for a new

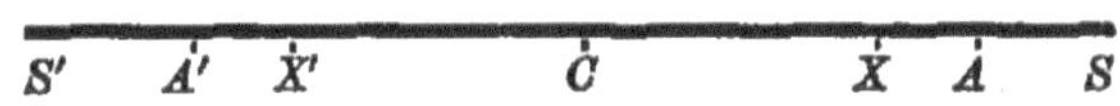

figure, the proof of these relations is identical with the proof given for the ellipse in Art. 61.

99. Prop. IV. *If S be a focus, and B an extremity of the conjugate axis,*

$$BC^2 = AS \,.\, SA', \text{ and } SC^2 = AC^2 + BC^2.$$

Referring to Art. (98), $SX = SA + AX$;

$$\therefore SX : AX :: SA + AX : AX,$$
$$:: SC + AC : AC;$$

and similarly

$$SX : A'X :: SC - AC : AC;$$
$$\therefore SX^2 : AX \,.\, A'X :: SC^2 - AC^2 : AC^2.$$

But

$$BC^2 : AC^2 :: SX^2 : AX \,.\, A'X;$$
$$\therefore BC^2 = SC^2 - AC^2 = AS \,.\, SA'.$$

Hence

$$SC^2 = AC^2 + BC^2 = AB^2;$$

i.e. SC is equal to the line joining the ends of the axes.

100. Prop. V. *The difference of the focal distances of any point is equal to the transverse axis.*

For, if PKK', perpendicular to the directrices, meet them in K and K',

$$S'P : PK' :: SA : AX,$$

and

$$SP : PK :: SA : AX;$$

$$\therefore S'P - SP : KK' :: SA : AX,$$

$$:: AA' : XX' \text{ (Art. 98)};$$

$$\therefore S'P - SP = AA'.$$

Cor. 1.

$$SP : NX :: AC : CX;$$

$$\therefore SP : AC :: NX : CX;$$

$$\therefore SP + AC : AC :: CN : CX,$$

or

$$SP + AC : CN :: SA : AX.$$

Hence also

$$S'P - AC : CN :: SA : AX.$$

Cor. 2. *Hence also it can be easily shewn, that the difference of the distances of any point from the foci of an hyperbola, is greater or less than the transverse axis, according as the point is within or without the concave side of the curve.*

101. *Mechanical Construction of the Hyperbola.*

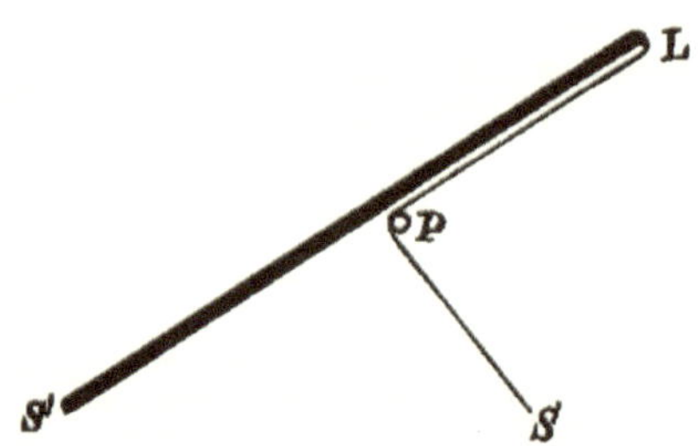

Let a straight rod $S'L$ be moveable in the plane of the paper about the point S'. Take a piece of string, the length of which is less than that of the rod, and fasten one end to a fixed point S, and the other end to L; then, pressing a pencil against the string so as to keep it stretched, and a part of it PL in contact with the rod, the pencil will trace out on the paper an hyperbola, having its foci at S and S', and its transverse axis equal to the difference between the length of the rod and that of the string.

This construction gives the right-hand branch of the curve; to trace the other branch, take the string longer than the rod, and such that it exceeds the length of the rod by the transverse axis.

We may remark that by taking a longer rod $MS'L$, and taking the string longer than $SS' + S'L$, so that the point P will be always on the end $S'M$ of the rod, we shall obtain an ellipse of which S and S' are the foci. Moreover, remembering that a parabola is the limiting form of an ellipse when one of the foci is removed to an infinite distance, the mechanical construction given for the parabola will be seen to be a particular case of the above.

The Asymptotes.

102. We have shewn in Art. 94 that if two points, e and e', be taken on the directrix such that

$$Ae = Ae' = AS,$$

the lines eA, $e'A$ meet the curve at an infinite distance.

These lines are parallel to the diagonals of the rectangle formed by the axes, for

$$Ae' : AX :: AS : AX :: SC : AC,$$
$$:: AB : AC, \text{ (Art. 99)}.$$

Definition. *The diagonals of the rectangle formed by the principal axes are called the asymptotes.*

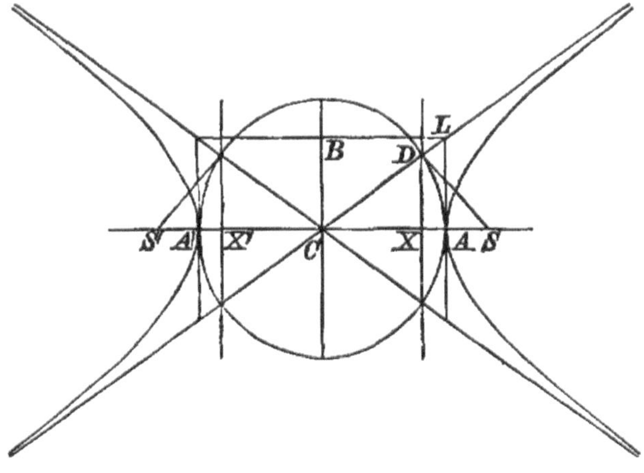

We observe that the axes bisect the angles between the asymptotes, and that if a double ordinate, PNP', when produced, meet the asymptotes in Q and Q',

$$PQ = P'Q'.$$

The figure appended will give the general form of the curve and its connection with the asymptotes and the auxiliary circle.

103. PROP. VI. *The asymptotes intersect the directrices in the same points as the auxiliary circle, and the lines joining the corresponding foci with the points of intersection are tangents to the circle.*

If the asymptote CL meet the directrix in D, joining SD (fig. Art. 102), $CL^2 = AC^2 + BC^2 = SC^2$,

and $$CD : CX :: CL : CA :: SC : CA :: CA : CX;$$

$\therefore CD = CA$, and D is on the auxiliary circle.

Also

$$CS \,.\, CX = CA^2 = CD^2;$$

$\therefore CDS$ is a right angle, and SD is the tangent at D.

COR. $CD^2 + SD^2 = CS^2 = AC^2 + BC^2$ (Art. 99);

$$\therefore SD = BC.$$

104. An asymptote may also be characterized as the ultimate position of a tangent when the point of contact is removed to an infinite distance.

It appears from Art. 10 that in order to find the point of contact of a tangent drawn from a point T in the directrix, we must join T with the focus S, and draw through S a straight line at right angles to ST; this line will meet the curve in the point of contact.

In the figures of Arts. 94 and 102 we know that the line through S, parallel to eA or CL, meets the curve in a point at an infinite distance, and also that this straight line is at right angles to SD, since SD is at right angles to CD. Hence the tangent from D, that is the line from D to the point at an infinite distance, is perpendicular to DS and therefore coincident with CD.

The asymptotes therefore touch the curve at an infinite distance.

105. DEF. *If an hyperbola be described, having for its transverse and conjugate axes, respectively, the conjugate and transverse axes of a given hyperbola, it is called the conjugate hyperbola.*

It is evident from the preceding article that the conjugate hyperbola has the same asymptotes as the original hyperbola, and that the distances of its foci from the centre are also the same.

The relations of Art. 96 and its Corollary are also true, *mutatis mutandis*, of the conjugate hyperbola; thus, if R be a point in the conjugate hyperbola,

$$RM^2 : CM^2 - BC^2 :: AC^2 : BC^2,$$

and

$$CM^2 : RM^2 + AC^2 :: BC^2 : AC^2.$$

DEF. *A straight line drawn through the centre and terminated by the conjugate hyperbola is also called a diameter of the original hyperbola.*

106. PROP. VII. *If from any point Q in one of the asymptotes, two straight lines QPN, QRM be drawn at right angles respectively to the transverse and conjugate axes, and meeting the hyperbola in P, p, and the conjugate hyperbola in R, r,*

$$QP \,.\, Qp = BC^2,$$

and

$$QR \,.\, Qr = AC^2.$$

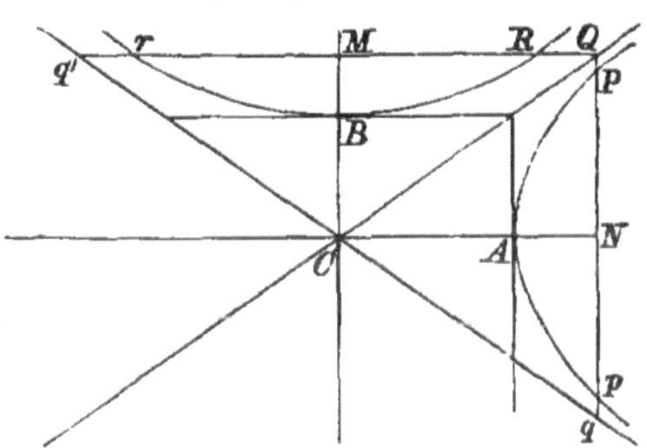

For

$$QN^2 : BC^2 :: CN^2 : AC^2;$$

$$\therefore QN^2 - BC^2 : BC^2 :: CN^2 - AC^2 : AC^2$$

$$:: PN^2 : BC^2;$$

$$\therefore QN^2 - BC^2 = PN^2,$$

or

$$QN^2 - PN^2 = BC^2;$$

i.e.

$$QP \,.\, Qp = BC^2.$$

Similarly,

$$QM^2 : AC^2 :: CM^2 : BC^2;$$

$$\therefore QM^2 - AC^2 : AC^2 :: CM^2 - BC^2 : BC^2,$$

$$:: RM^2 : AC^2;$$

$$\therefore QM^2 - RM^2 = AC^2,$$

or

$$QR \,.\, Qr = AC^2.$$

These relations may also be given in the form,

$$QP \,.\, Pq = BC^2, \quad QR \,.\, Rq' = AC^2.$$

Cor. If the point Q be taken at a greater distance from C, the length QN and therefore Qp will be increased, and may be increased indefinitely.

But the rectangle $QP \,.\, Qp$ is of finite magnitude; hence QP will be indefinitely diminished, and the curve, therefore, as it recedes from the centre, tends more and more nearly to coincide with the asymptote.

A further illustration is thus given of the remarks in Art. 104.

107. If in the preceding figure the line Qq be produced to meet the conjugate hyperbola in E and e, it can be shewn, in the same manner as in Art. 106, that

$$QE \,.\, Qe = BC^2;$$

and this equality is still true when the line Qq lies between C and A, in which case Qq does not meet the hyperbola.

Properties of the Tangent and Normal.

108. In the case of the hyperbola the theorem, proofs of which are given in Arts. 15 and 16, takes the following form:

The tangents drawn from any point to an hyperbola subtend equal or supplementary angles at either focus according as they touch the same or opposite branches of the curve.

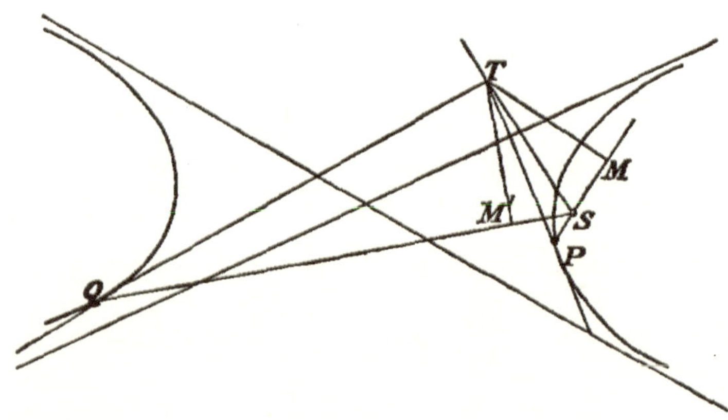

For, T being the point of intersection of tangents to opposite branches of the curve, let TM, TM' be the perpendiculars let fall from T on SP and SQ, then, as in Arts. 15 and 16, $TM = TM'$;

$\therefore$ the angles TSM, TSM' are equal, and consequently the angles TSP, TSQ are supplementary.

109. PROP. VIII. *The tangent at any point bisects the angle between the focal distances of that point, and the normal is equally inclined to the focal distances.*

Let the normal at P meet the axis in G.

Then (Art. 18),

$$SG : SP :: SA : AX,$$

and
$$S'G : S'P :: SA : AX;$$

$$\therefore SG : S'G :: SP : S'P;$$

and therefore the angle between SP and $S'P$ produced is bisected by PG.

Hence PT, the tangent which is perpendicular to PG, bisects the angle SPS'.

COR. 1. If PT and GP produced meet, respectively, the conjugate axis in t and g, it can be shewn, in exactly the same manner as in the corresponding case of the ellipse (Art. 68), that S, P, S', t, and g are concyclic.

COR. 2. If an ellipse be described having S and S' for its foci, and if this ellipse meet the hyperbola in P, the normal at P to the ellipse bisects the angle SPS', and therefore coincides with the tangent to the hyperbola.

Hence, if an ellipse and an hyperbola be confocal, that is, have the same foci, they intersect at right angles.

110. PROP. IX. *Every diameter is bisected at the centre, and the tangents at the ends of a diameter are parallel.*

Let PCp be a diameter, and PN, pn the ordinates.

Then
$$CN^2 : Cn^2 :: PN^2 : pn^2,$$
$$:: CN^2 - AC^2 : Cn^2 - AC^2;$$

hence $CN = Cn$, and $\therefore CP = Cp$.

Again, if PT, pt be the tangents,

The triangles PCS, pCS' are equal in all respects, and therefore $SPS'p$ is a parallelogram.

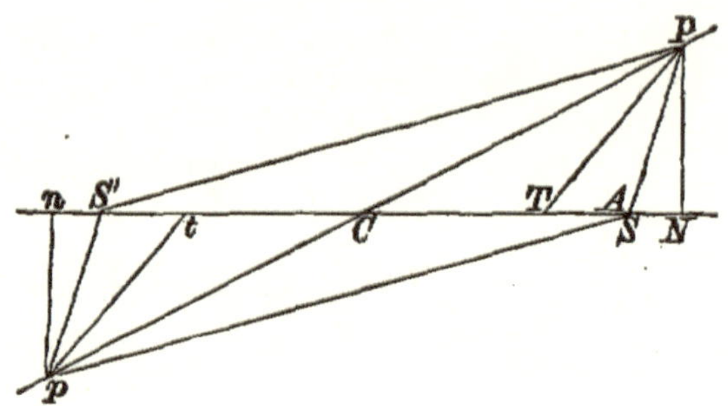

Hence the angles SPS', SpS' are equal, and therefore $SPT = S'pt$.

But $SPC = S'pC$,

$\therefore$ the difference TPC = the difference tpC, and PT is parallel to pt.

It can be shewn in exactly the same manner, that, if the diameter be terminated by the conjugate hyperbola, it is bisected in C, and the tangents at its extremities are parallel.

COR. The distances SP, Sp are equally inclined to the tangents at P and p.

111. PROP. X. *The perpendiculars from the foci on any tangent meet the tangent on the auxiliary circle, and the semi-conjugate axis is a mean proportional between their lengths.*

Let SY, $S'Y'$ be the perpendiculars, and let SY produced meet $S'P$ in L.

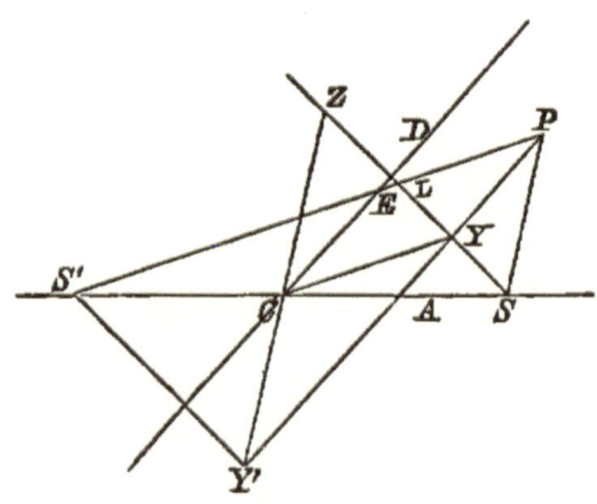

Then the triangles SPY, LPY are equal in all respects,

$$\text{and } SY = LY.$$

Hence, C being the middle point of SS' and Y of SL, CY is parallel to $S'L$, and $S'L = 2CY$.

But
$$S'L = S'P - PL = S'P - SP = 2AC;$$
$$\therefore CY = AC,$$

and Y is on the auxiliary circle.

So also Y' is a point in the circle.

Let SY produced meet the circle in Z, and join $Y'Z$; then, $Y'YZ$ being a right angle, ZY' is a diameter and passes through C. Hence, the triangles SCZ, $S'CY'$ being equal,

$$S'Y' = SZ,$$

and
$$SY \,.\, S'Y' = SY \,.\, SZ = SA \,.\, SA' = BC^2.$$

Cor. 1. If P' be the other extremity of the diameter PC, the tangent at P' is parallel to PY, and therefore Z is the foot of the perpendicular from S on the tangent at P'.

Cor. 2. If the diameter DCD', drawn parallel to the tangent at P, meet $S'P$, SP in E and E', $PECY$ is a parallelogram;

$$\therefore PE = CY = AC,$$

and so also
$$PE' = CY' = AC.$$

112. Prop. XI. *To draw tangents to an hyperbola from a given point.*

The construction of Art. 17 may be employed, or, as in the cases of the ellipse and parabola, the following.

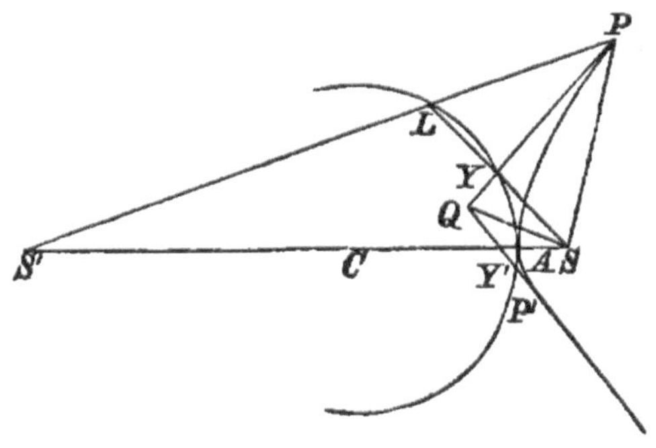

Let Q be the given point; join SQ, and upon SQ as diameter describe a circle intersecting the auxiliary circle in Y and Y';

QY and QY' are the required tangents.

Producing SY to L, so that $YL = SY$, draw $S'L$ cutting QY in P, and join SP.

The triangles SPY, LPY are equal in all respects,

and $$S'P - SP = S'L = 2CY = 2AC;$$

$\therefore P$ is a point on the hyperbola.

Also QP bisects the angle SPS', and is therefore the tangent at P. A similar construction will give the other tangent QP'.

If the point Q be within the angle formed by the asymptotes, the tangents will both touch the same branch of the curve; but if it lie within the external angle, they will touch opposite branches.

113. PROP. XII. *If two tangents be drawn from any point to an hyperbola they are equally inclined to the focal distances of that point.*

Let PQ, $P'Q$ be the tangents, SY, $S'Y'$, SZ, $S'Z'$ the perpendiculars from the foci; join YZ, $Y'Z'$.

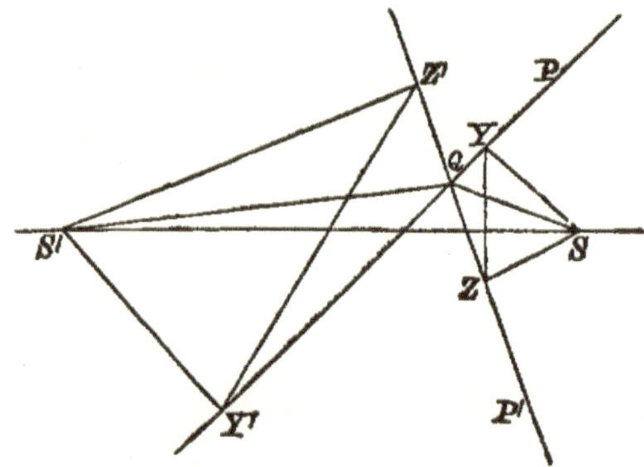

Then the angles YSZ, $Y'S'Z'$ are equal, for they are the supplements of YQZ, $Y'QZ'$.

Also $$SY \,.\, S'Y' = SZ \,.\, S'Z' \text{ (Art. 111)};$$

or $$SY : SZ :: S'Z' : S'Y';$$

$$\therefore \text{ the triangles } YZS,\ Y'S'Z' \text{ are similar,}$$
$$\text{and the angle } YZS = Z'Y'S'.$$
$$\text{But the angle } YQS = YZS, \text{ and } Z'QS' = ZY'S';$$
$$\therefore YQS = Z'QS'.$$

That is, the tangent QP and the tangent $P'Q$ produced are equally inclined to SQ and $S'Q$.

Or, producing $S'Q$, QP and QP' are equally inclined to QS and $S'Q$ produced.

In exactly the same manner it can be shewn that if QP, QP' touch opposite branches of the curve the angles PQS, $P'QS'$ are equal.

Cor. If Q be a point in a confocal hyperbola, the normal at Q bisects the angle between SQ and $S'Q$ produced and therefore bisects the angle PQP'.

Hence, if from any point of an hyperbola tangents be drawn to a confocal hyperbola, these tangents are equally inclined to the normal or the tangent at the point, according as it lies within or without that angle formed by the asymptotes of the confocal which contains the transverse axes.

114. Prop. XIII. *If PT, the tangent at P, meet the transverse axis in T, and PN be the ordinate,*

$$CN \,.\, CT = AC^2$$

Let fall the perpendicular SY upon PT, and join YN, CY, SP, and $S'P$.

The angle $\quad CYT = S'PY = SPY$

$$= \text{the supplement of } SNY = CNY;$$

also the angle YCT is common to the two triangles CYT, CYN; these triangles are therefore similar, and

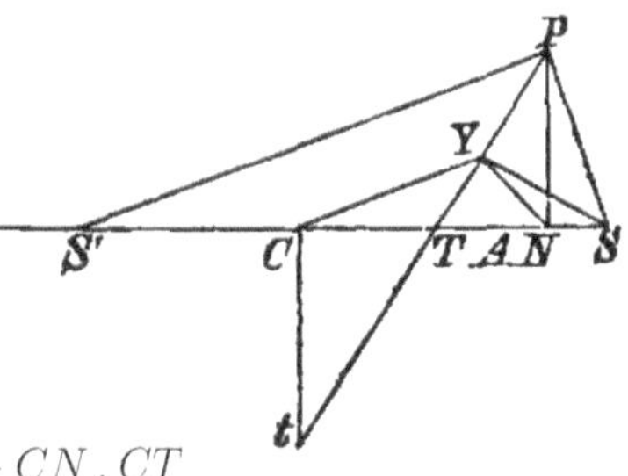

$$CN : CY :: CY : CT,$$

or

$$CN \,.\, CT = CY^2 = AC^2$$

Cor. 1. Hence $\quad CN \,.\, NT = CN^2 - CN \,.\, CT$

$$= CN^2 - AC^2$$
$$= AN \,.\, NA'$$

COR. 2. Hence also it follows that

If any number of hyperbolas be described having the same transverse axis, and an ordinate be drawn cutting the hyperbolas, the tangents at the points of section will all meet the transverse axis in the same point.

COR. 3. If CN be increased indefinitely, CT is diminished indefinitely, and the tangent ultimately passes through C, as we have already shewn in Art. 104.

115. PROP. XIV. *If the tangent at P meet the conjugate axis in t, and PN be the ordinate,*

$$Ct \,.\, PN = BC^2.$$

For $$Ct : PN :: CT : NT; \text{ (Fig. Art. 114)}$$

$$\therefore Ct \,.\, PN : PN^2 :: CT \,.\, CN : CN \,.\, NT$$
$$:: AC^2 : AN \,.\, NA'.$$
$$\therefore Ct \,.\, PN : AC^2 :: PN^2 : AN \,.\, NA'$$
$$:: BC^2 : AC^2,$$

and $$Ct \,.\, PN = BC^2.$$

In exactly the same manner as in Art. 76, it can be shewn that

$$CG \,.\, CT = SC^2,$$

$$CG : CN :: SC^2 : AC^2, \quad Cg : PN :: SC^2 : BC^2,$$

and $$NG : CN :: BC^2 : AC^2.$$

116. PROP. XV. *If the normal at P meet the transverse axis in G, the conjugate axis in g, and the diameter parallel to the tangent at P in F,*

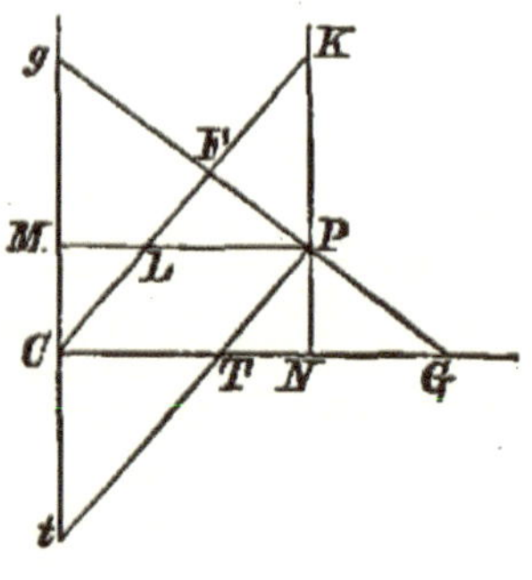

$$PF \,.\, PG = BC^2, \text{ and } PF \,.\, Pg = AC^2.$$

Let NP, PM, perpendicular to the axes, meet the diameter CF in K and L;

Then KNG, KFG being right angles, K, F, N, G are concyclic;

$$\therefore PF \,.\, PG = PK \,.\, PN$$
$$= Ct \,.\, PN = BC^2.$$

Similarly F, L, M, g are concyclic;

$$\therefore PF \,.\, Pg = PL \,.\, PM = CT \,.\, CN = AC^2.$$

117. Prop. XVI. *If PCp be a diameter, and QV an ordinate, and if the tangent at Q meet the diameter Pp in T,*

$$CV \,.\, CT = CP^2.$$

Let the tangents at P and p meet the tangent at Q in R and r;

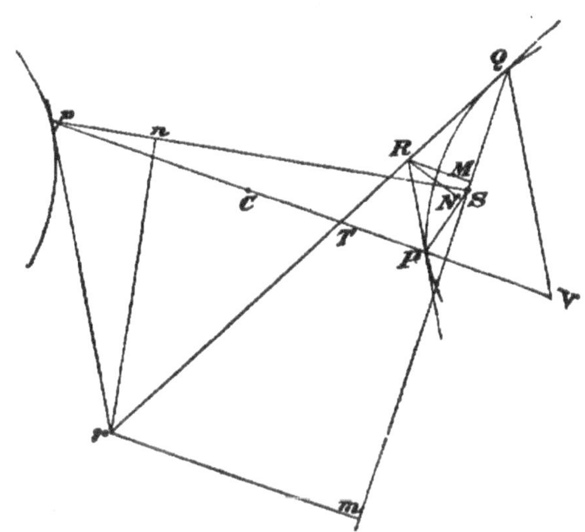

Then the angle $SPR = Spr$ (Cor. Art. 110)
and therefore if RN, rn be the perpendiculars on SP, sp, the triangles RPN, rpn are similar.

Draw RM, rm perpendiculars on SQ.

Then $\quad TR : Tr :: RP : rp :: RN : rn,$

$$:: RM : rm \text{ (Cor. Art. 15)}$$

$$:: RQ : rQ.$$

Hence, QV, RP, and rp being parallel,

$$TP : Tp :: PV : pV;$$

$$\therefore TP + Tp : Tp - TP :: PV + pV : pV - PV,$$

or
$$2CP : 2CT :: 2CV : 2CP,$$

or
$$CV \,.\, CT = CP^2.$$

118. Prop. XVII. *A diameter bisects all chords parallel to the tangents at its extremities.*

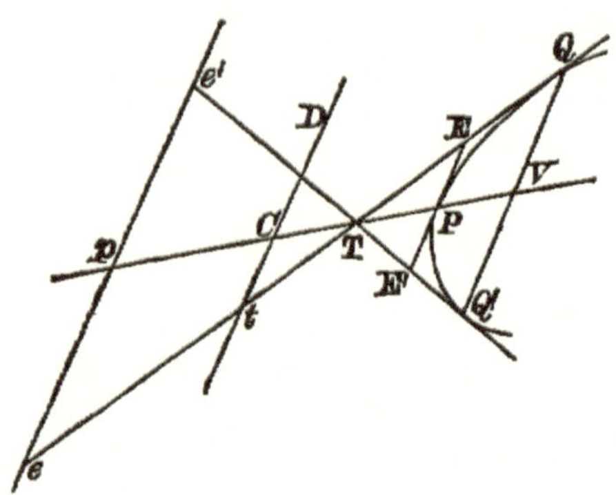

Let PCp be the diameter, and QQ' the chord, parallel to the tangents at P and p. Then if the tangents TQ, TQ' at Q and Q' meet the tangents at P and p, in the points E, E', e, e',

$$EP = E'P \text{ and } ep = e'p, \text{ (Art. 21)}$$

$$\therefore \text{the point } T \text{ is on the line } Pp;$$

$$\text{but } TP \text{ bisects } QQ';$$

that is, the diameter pCP produced bisects QQ'.

Def. *The line DCd, drawn parallel to the tangent at P and terminated by the conjugate hyperbola, that is, the diameter parallel to the tangent at P, is said to be conjugate to PCp.*

A diameter therefore bisects all chords parallel to its conjugate.

119. Prop. XVIII. *If the diameter DCd be conjugate to PCp, then will PCp be conjugate to DCd.*

Let the chord QVq be parallel to CD and be bisected in V by CP produced.

Draw the diameter qCR, and join RQ meeting CD in U.

Then $RC = Cq$ and $QV = Vq$; $\therefore QR$ is parallel to CP.

Also $$QU : UR :: Cq : CR,$$

and $$\therefore QU = UR,$$

that is, CD bisects the chords parallel to CP, and PCp is therefore conjugate to DCd.

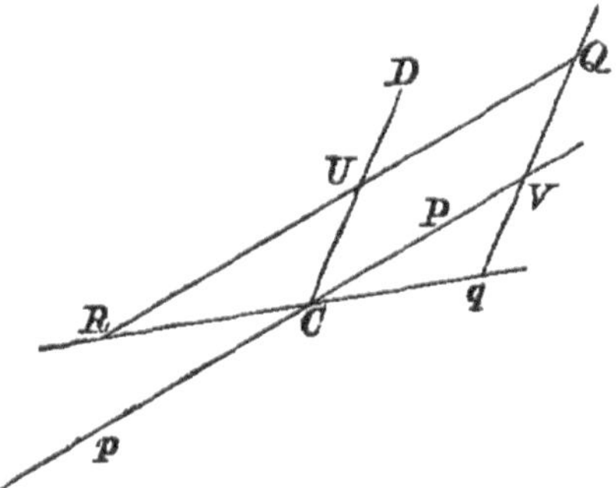

Hence, when two diameters are conjugate, each bisects the chords parallel to the other.

DEF. *Chords drawn from the extremities of any diameter to a point on the hyperbola are called supplemental chords.*

Thus, qQ, QR are supplemental chords, and they are parallel to CD and CP; supplemental chords are therefore parallel to conjugate diameters.

DEF. *A line QV, drawn from any point Q of an hyperbola, parallel to a diameter DCd, and terminated by the conjugate diameter PCp, is called an ordinate of the diameter PCp, and if QV produced meet the curve in Q', QVQ' is the double ordinate.*

This definition includes the two cases in which QQ' may be drawn so as to meet the same, or opposite branches of the hyperbola.

120. PROP. XIX. *Any diameter is a mean proportional between the transverse axis and the focal chord parallel to the diameter.*

This can be proved as in Art. 81.

Properties of Asymptotes.

121. PROP. XX. *If from any point Q in an asymptote $QPpq$ be drawn meeting the curve in P, p and the other asymptote in q, and if CD be the semi-diameter parallel to Qq,*

$$QP \,.\, Pq = CD^2 \text{ and } QP = pq.$$

Through P and D draw RPr, DTt perpendicular to the transverse axis, and meeting the asymptotes.

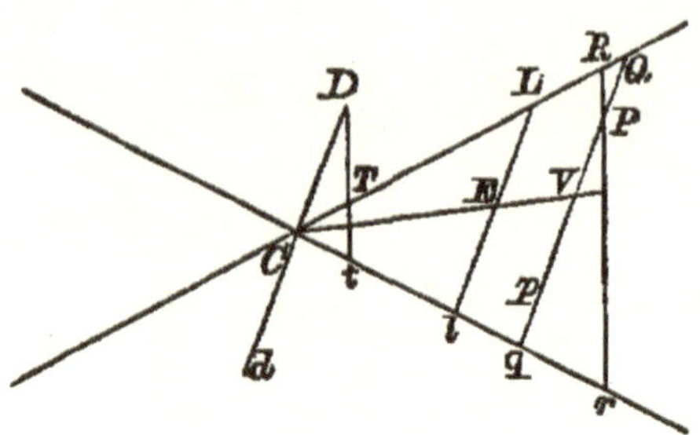

Then $$QP : RP :: CD : DT,$$

and $$Pq : Pr :: CD : Dt;$$

$$\therefore QP \,.\, Pq : RP \,.\, Pr :: CD^2 : DT \,.\, Dt.$$

But $$RP \,.\, Pr = BC^2 = DT \,.\, Dt \text{ (Arts. 106 and 107)},$$

$$\therefore QP \,.\, Pq = CD^2.$$

Similarly $$qp \cdot pQ = CD^2;$$

$$\therefore QP \,.\, Pq = qp \,.\, pQ;$$

or, if V be the middle point of Qq,

$$QV^2 - PV^2 = QV^2 - pV^2.$$

Hence $$PV = pV, \text{ and } \therefore PQ = pq.$$

We have taken the case in which Qq meets one branch of the hyperbola. It may however be shewn in the same manner that the same relations hold good for the case in which Qq meets opposite branches.

COR. *If a straight line $PP'p'p$ meet the hyperbola in P, p, and the conjugate hyperbola in P', p', $PP' = pp'$.*

For, if the line meet the asymptotes in Q, q,

$$QP' = p'q, \text{ and } PQ = qp;$$

$$\therefore PP' = pp'.$$

122. PROP. XXI. *The portion of a tangent which is terminated by the asymptotes is bisected at the point of contact, and is equal to the parallel diameter.*

LEl being the tangent (Fig. Art. 121), and DCd the parallel diameter, draw any parallel straight line $QPpq$ meeting the curve and the asymptotes.

Then $QP = pq$; and, if the line move parallel to itself until it coincides with Ll, the points P and p coincide with E, and $\therefore LE = El$.

Also $$QP \,.\, Pq = CD^2, \text{ always;}$$
$$\therefore LE \,.\, El = CD^2, \text{ or } LE = CD.$$

Properties of Conjugate Diameters.

123. Prop. XXII. *Conjugate diameters of an hyperbola are also conjugate diameters of the conjugate hyperbola, and the asymptotes are diagonals of the parallelogram formed by the tangents at their extremities.*

PCp and DCd being conjugate, let QVq, a double ordinate of CD, meet the conjugate hyperbola in Q' and q'.

Then $$QV = Vq, \text{ and } QQ' = qq' \text{ (Cor. Art. 121)},$$
$$\therefore Q'V = Vq'.$$

That is, CD bisects the chords of the conjugate hyperbola parallel to CP.

Hence CD and CP are conjugate in both hyperbolas, and therefore the tangent at D is parallel to CP.

Let the tangent at P meet the asymptote in L; then
$$PL = CD \text{ (Art. 122)}.$$

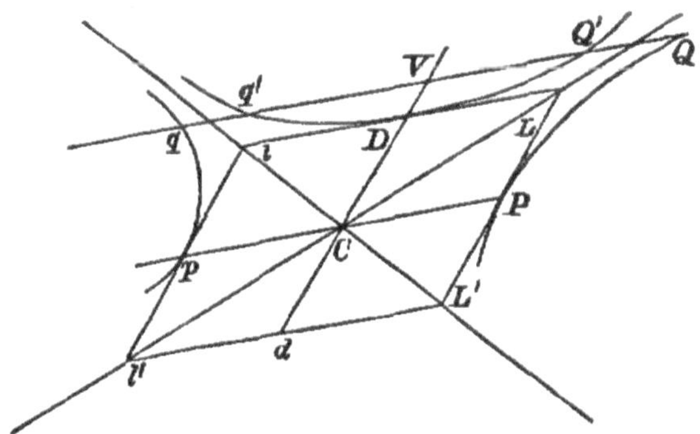

Hence LD is parallel and equal to CP;
but the tangent at D is parallel to CP;
$$\therefore LD \text{ is the tangent at } D.$$

Completing the figure, the tangents at p and d are parallel to those at P and D, and therefore the asymptotes are the diagonals of the parallelogram $Ll'l'L'$.

COR. Hence, joining PD, it follows that PD is parallel to the asymptote lCL', since $LP = PL'$, and $LD = Dl$,

124. PROP. XXIII. *If QV be an ordinate of a diameter PCp, and DCd the conjugate diameter,*

$$QV^2 : PV \, . \, Vp :: CD^2 : CP^2.$$

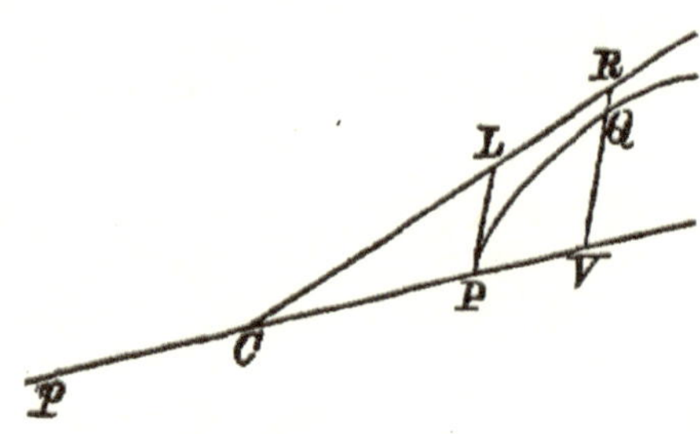

Let QV and the tangent at P meet the asymptote in R and L.

Then LP being equal to CD,

$$RV^2 : CD^2 :: CV^2 : CP^2;$$

$$\therefore RV^2 - CD^2 : CD^2 :: CV^2 - CP^2 : CP^2.$$

But $$RV^2 - QV^2 = CD^2.$$

Hence $$QV^2 : CD^2 :: CV^2 - CP^2 : CP^2,$$

$$\text{or } QV^2 : PV \, . \, Vp :: CD^2 : CP^2.$$

125. PROP. XXIV. *If QV be an ordinate of a diameter PCp, and if the tangent at Q meet the conjugate diameter, DCd, in t,*

$$Ct \, . \, QV = CD^2.$$

For, (Fig. Art. 118)

$$Ct : QV :: CT : VT,$$

$$\text{and } \therefore Ct \, . \, QV : QV^2 :: CV \, . \, CT : CV \, . \, VT.$$

But $$CV \, . \, CT = CP^2,$$

$$\text{and } CV \, . \, VT = CV^2 - CV \, . \, CT = CV^2 - CP^2;$$

$$\therefore Ct \, . \, QV : QV^2 :: CP^2 : CV^2 - CP^2,$$

$$:: CD^2 : QV^2.$$

Hence $$Ct \, . \, QV = CD^2.$$

126. Prop. XXV. *If ACa, BCb be conjugate diameters, and PCp, DCd another pair of conjugate diameters, and if PN, DM be ordinates of ACa,*

$$CM : PN :: AC : BC,$$
$$\text{and } DM : CN :: BC : AC.$$

Let the tangents at P and D meet ACa in T and t;
then $CN \,.\, CT = AC^2 = CM \,.\, Ct$ (Art. 117),

$$\begin{aligned} \therefore CM : CN &:: CT : Ct, \\ &:: PT : CD, \\ &:: PN : DM, \\ &:: CN : Mt; \end{aligned}$$

$$\therefore CN^2 = CM \,.\, Mt = CM^2 + CM \,.\, Ct = CM^2 + AC^2,$$

so that $$CM^2 = CN^2 - AC^2.$$

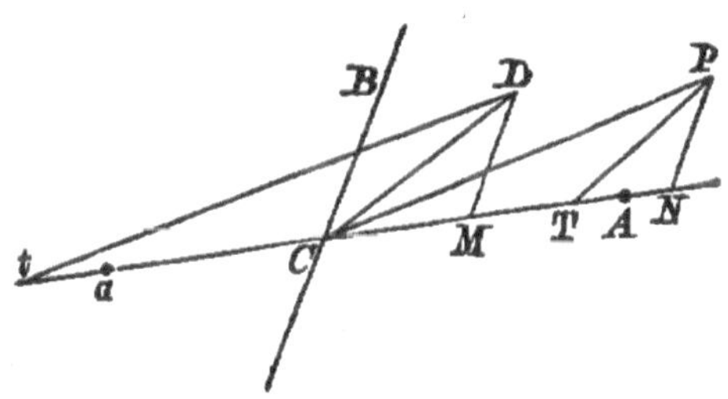

But $$PN^2 : CN^2 - AC^2 :: BC^2 : AC^2;$$
$$\therefore CM : PN :: AC : BC;$$
and, similarly, $$DM : CN :: BC : AC.$$

Cor. We have shewn in the course of the proof, that

$$CN^2 - CM^2 = AC^2.$$

Similarly, if Pn, Dm be ordinates of BC,

$$Cm^2 - Cn^2 = BC^2;$$

that is, $$DM^2 - PN^2 = BC^2;$$

and it must be noticed that these relations are shewn for any pair of conjugate diameters ACa, BCb, including of course the axes.

127. **Prop. XXVI.** *If CP, CD be conjugate semi-diameters, and AC, BC the semi-axes,*

$$CP^2 - CD^2 = AC^2 - BC^2.$$

For, drawing the ordinates PN, DM, and remembering that in this case the angles at N and M are right angles, we have, from the figure of the previous article,

$$CP^2 = CN^2 + PN^2,$$
$$CD^2 = CM^2 + DM^2.$$

But $CN^2 - CM^2 = AC^2$ and $DM^2 - PN^2 = BC^2$;

$$\therefore CP^2 - CD^2 = AC^2 - BC^2.$$

128. **Prop. XXVII.** *If the normal at P meet the axes in G and g,*

$$PG : CD :: BC : AC,$$

and

$$Pg : CD :: AC : BC.$$

For the proofs of these relations, see Art. 86.

Observe also that

$$PG \,.\, Pg = CD^2,$$

and that

$$Gg : CD :: SC^2 : AC \,.\, BC.$$

129. **Prop. XXVIII.** *The area of the parallelogram formed by the tangents at the ends of conjugate diameters is equal to the rectangle contained by the axes.*

Let CP, CD be the semi-diameters, and PN, DM the ordinates of the transverse axis.

Let the normal at P meet CD in F, and the axis in G. Then PNG, CDM are similar triangles, and, exactly as in Art. 87, it can be shewn that

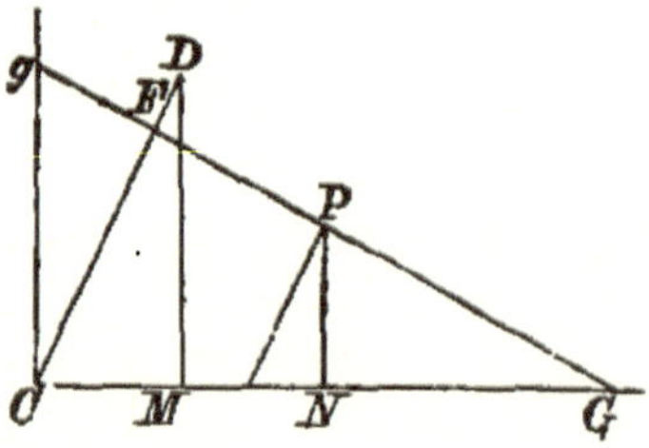

$$PF \,.\, CD = AC \,.\, BC.$$

Hence it follows that, in the figure of Art. 123, the triangle LCL' is of constant area.

For the triangle is equal to the parallelogram $CPLD$.

130. Prop. XXIX. *If SP, $S'P$ be the focal distances of a point P, and CD be conjugate to CP,*

$$SP \,.\, S'P = CD^2.$$

Attending to the figure of Art. 111, the proof is the same as that of Art. 88.

131. Prop. XXX. *If the tangent at P meet a pair of conjugate diameters in T and t, and CD be conjugate to CP,*

$$PT \,.\, Pt = CD^2.$$

This can be proved as in Art. 89.

It can also be shewn that if the tangent at P meet two parallel tangents in T' and t',

$$PT' \,.\, Pt' = CD^2.$$

132. Prop. XXXI. *If the tangent at P meet the asymptotes in L and L',*

$$CL \,.\, CL' = SC^2.$$

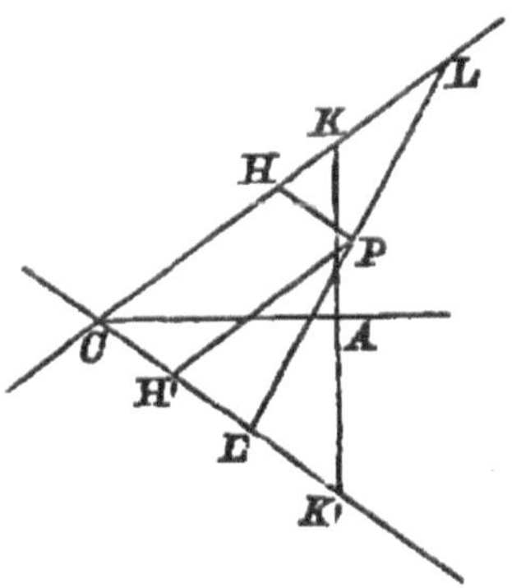

Let the tangent at A meet the asymptotes in K and K'; then (Art. 129) the triangles LCL', KCK' are of equal area, and therefore

$$CL : CK' :: CK : CL' \text{ (Euclid, Book VI.)},$$

or

$$CL \,.\, CL' = CK^2 = AC^2 + BC^2 = SC^2.$$

Cor. If PH, PH' be drawn parallel to, and terminated by the asymptotes,

$$4 \,.\, PH \,.\, PH' = CS^2,$$

for $CL = 2PH'$, and $CL' = 2PH$.

133. Prop. XXXII. *Pairs of tangents at right angles to each other intersect on a fixed circle.*

PT, QT being two tangents at right angles, let SY, perpendicular to PT, meet $S'P$ in K.

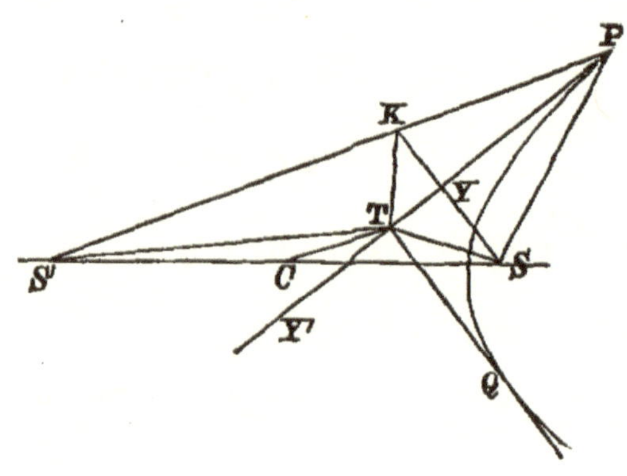

Then (Art. 113) the angle $S'TY' = QTS$,

and obviously, $KTP = PTS$;

therefore $S'TY'$ is complementary to KTP, and $S'TK$ is a right angle.

Hence

$$\begin{aligned} 4AC^2 = S'K^2 &= S'T^2 + TK^2 \\ &= S'T^2 + ST^2 \\ &= 2 \,.\, CT^2 + 2 \,.\, CS^2 \text{ by Euclid II. 12 and 13;} \end{aligned}$$

$$\therefore CT^2 = AC^2 - BC^2,$$

and the locus of T is a circle.

If AC be less than BC, this relation is impossible.

In this case, however, the angle between the asymptotes is greater than a right angle, and the angle PTQ between a pair of tangents, being always greater than the angle between the asymptotes, is greater than a right angle. The problem is therefore *à priori* impossible for the hyperbola, but becomes possible for the conjugate hyperbola.

As in the case of the ellipse, the locus of T is called the director circle.

134. Prop. XXXIII. *The rectangles contained by the segments of any two chords which intersect each other are in the ratio of the squares on the parallel diameters.*

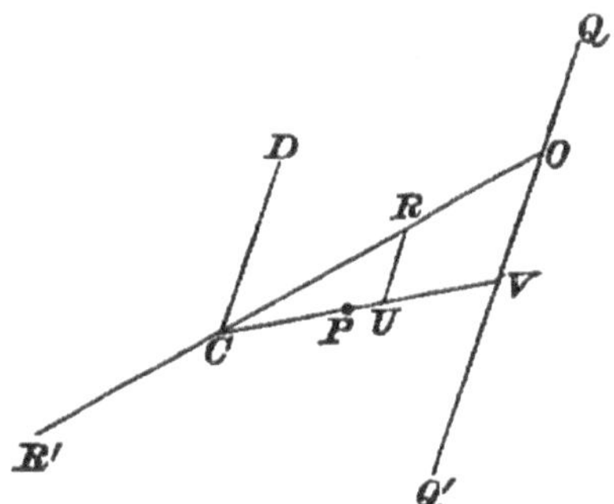

Through any point O in a chord QOQ' draw the diameter ORR'; and let CD be parallel to QQ', and CP conjugate to CD, bisecting QQ' in V.

Draw RU an ordinate of CP.

Then $$RU^2 : CU^2 - CP^2 :: CD^2 : CP^2;$$
$$\therefore CD^2 + RU^2 : CU^2 :: CD^2 : CP^2,$$
$$:: CD^2 + QV^2 : CV^2.$$

But $$RU^2 : CU^2 :: OV^2 : CV^2;$$

$$\therefore CD^2 : CU^2 :: CD^2 + QV^2 - OV^2 : CV^2,$$

or $$CD^2 : CD^2 + QV^2 - OV^2 :: CU^2 : CV^2,$$
$$:: CR^2 : CO^2;$$

$$\therefore CD^2 : QV^2 - OV^2 :: CR^2 : CO^2 - CR^2,$$

or $$CD^2 : QO \,.\, OQ' :: CR^2 : OR \,.\, OR'.$$

Similarly, if qOq' be any other chord, and Cd the parallel semi-diameter,
$$Cd^2 : qO \,.\, Oq' :: CR^2 : OR \,.\, OR';$$
$$\therefore QO \,.\, OQ' : qO \,.\, Oq' :: CD^2 : Cd^2;$$
that is, the ratio of the rectangles depends only on the directions of the chords.

PROP. XXXIV. *If a circle intersect an hyperbola in four points, the several pairs of the chords of intersection are equally inclined to the axes.*

For the proof, see Art. 93.

EXAMPLES.

1. If a circle be drawn so as to touch two fixed circles externally, the locus of its centre is an hyperbola.

2. If the tangent at B to the conjugate meet the latus rectum in D, the triangles SCD, SXD are similar.

3. The straight line drawn from the focus to the directrix, parallel to an asymptote, is equal to the semi-latus rectum, and is bisected by the curve.

4. Given the asymptotes and a focus, find the directrix.

5. Given the centre, one asymptote, and a directrix, find the focus.

6. Parabolas are described passing through two fixed points, and having their axes parallel to a fixed line; the locus of their foci is an hyperbola.

7. The base of a triangle being given, and also the point of contact with the base of the inscribed circle, the locus of the vertex is an hyperbola.

8. If the normal at P meet the conjugate axis in g, and gN be the perpendicular on SP, then $PN = AC$.

9. Draw a tangent to an hyperbola, or its conjugate, parallel to a given line.

10. If AA' be the axis of an ellipse, and PNP' a double ordinate, the locus of the intersection of $A'P$ and $P'A$ is an hyperbola.

11. The tangent at P bisects any straight line perpendicular to AA', and terminated by AP, and $A'P$.

12. If PCp be a diameter, and if Sp meet the tangent at P in T,

$$SP = ST.$$

13. Given an asymptote, the focus, and a point; construct the hyperbola.

14. A circle can be drawn through the foci and the intersections of any tangent with the tangents at the vertices.

15. Given an asymptote, the directrix, and a point; construct the hyperbola.

16. If through any point of an hyperbola straight lines are drawn parallel to the asymptotes and meeting any semi-diameter CQ in P and R,

$$CP \,.\, CR = CQ^2.$$

17. PN is an ordinate and NQ parallel to AB meets the conjugate axis in Q; prove that $QB \,.\, QB' = PN^2$.

18. NP is an ordinate and Q a point in the curve; AQ, $A'Q$ meet NP in D and E; prove that $ND \,.\, NE = NP^2$.

19. If a tangent cut the major axis in the point T, and perpendiculars SY, HZ be let fall on it from the foci, then

$$AT \,.\, A'T = YT \,.\, ZT.$$

20. In the tangent at P a point Q is taken such that PQ is proportional to CD; shew that the locus of Q is an hyperbola.

21. Tangents are drawn to an hyperbola, and the portion of each tangent intercepted by the asymptotes is divided in a constant ratio; prove that the locus of the point of section is an hyperbola.

22. If the tangent and normal at P meet the conjugate axis in t and K respectively, prove that a circle can be drawn through the foci and the three points P, t, K.

Shew also that

$$GK : SK :: SA : AX,$$

and

$$St : tK :: BC : CD,$$

CD being conjugate to CP.

23. Shew that the points of trisection of a series of conterminous circular arcs lie on branches of two hyperbolas; and determine the distance between their centres.

24. If the tangent at any point P cut an asymptote in T, and if SP cut the same asymptote in Q, then $SQ = QT$.

25. A series of hyperbolas having the same asymptotes is cut by a straight line parallel to one of the asymptotes, and through the points of intersection lines are drawn parallel to the other, and equal to either semi-axis of the corresponding hyperbola: prove that the locus of their extremities is a parabola.

26. Prove that the rectangle $PY \,.\, PY'$ in an ellipse is equal to the square on the conjugate axis of the confocal hyperbola passing through P.

27. If the tangent at P meet one asymptote in T, and a line TQ be drawn parallel to the other asymptote to meet the curve in Q; prove that if PQ be joined and produced both ways to meet the asymptotes in R and R', RR' will be trisected at the points P and Q.

28. The tangent at a point P of an ellipse meets the hyperbola having the same axes as the ellipse in C and D. If Q be the middle point of CD, prove that OQ and OP are equally inclined to the axes, O being the centre of the ellipse.

29. Given one asymptote, the direction of the other, and the position of one focus, determine the position of the vertices.

30. Two points are taken on the same branch of the curve, and on the same side of the axis; prove that a circle can be drawn touching the four focal distances.

31. Supposing the two asymptotes and one point of the curve to be given in position, shew how to construct the curve; and find the position of the foci.

32. Given a pair of conjugate diameters, construct the axes.

33. If PH, PK be drawn parallel to the asymptotes from a point P on the curve, and if a line through the centre meet them in R, T, and the parallelogram $PRQT$ be completed, Q is a point on the curve.

34. The ordinate NP at any point of an ellipse is produced to a point Q, such that NQ is equal to the sub-tangent at P; prove that the locus of Q is an hyperbola.

35. If a given point be the focus of any hyperbola, passing through a given point and touching a given straight line, prove that the locus of the other focus is an arc of a fixed hyperbola.

36. An ellipse and hyperbola are described, so that the foci of each are at the extremities of the transverse axis of the other; prove that the tangents at their points of intersection meet the conjugate axis in points equidistant from the centre.

37. A circle is described about the focus as centre, with a radius equal to one-fourth of the latus rectum; prove that the focal distances of the points at which it intersects the hyperbola are parallel to the asymptotes.

38. The tangent at any point forms a triangle with the asymptotes: determine the locus of the point of intersection of the straight lines drawn from the angles of this triangle to bisect the opposite sides.

39. If SY, $S'Y'$ be the perpendiculars on the tangent at P, a circle can be drawn through the points Y, Y', N, C.

40. The straight lines joining each focus to the foot of the perpendicular from the other focus on the tangent meet on the normal and bisect it.

41. If the tangent and normal at P meet the axis in T and G, $NG.CT = BC^2$.

42. If the tangent at P meet the axes in T and t, the angles PSt, STP are supplementary.

43. If the tangent at P meet any conjugate diameters in T and t, the triangles SPT, $S'Pt$ are similar.

44. If the diameter conjugate to CP meet SP and $S'P$ in E and E', prove that the circles about the triangles SCE, $S'CE'$ are equal.

45. The locus of the centre of the circle inscribed in the triangle SPS' is a straight line.

46. If PN be an ordinate, and NQ parallel to AP meet CP in Q, AQ is parallel to the tangent at P.

47. If an asymptote meet the directrix in D, and the tangent at the vertex in E, AD is parallel to SE.

48. The radius of the circle touching the curve and its asymptotes is equal to the portion of the latus rectum produced, between its extremity and the asymptote.

49. If G be the foot of the normal, and if the tangent meet the asymptotes in L and M, $GL = GM$.

50. With two conjugate diameters of an ellipse as asymptotes, a pair of conjugate hyperbolas is constructed: prove that if one hyperbola touch the ellipse, the other will do so likewise; prove also that the diameters drawn through the points of contact are conjugate to each other.

51. If two tangents be drawn the lines joining their intersections with the asymptotes will be parallel.

52. The locus of the centre of the circle touching SP, $S'P$ produced, and the major axis, is an hyperbola.

53. If from a point P in an hyperbola, PK be drawn parallel to an asymptote to meet the directrix in K, then $PK = SP$.

54. If PD be drawn parallel to an asymptote, to meet the conjugate hyperbola in D, CP and CD are conjugate diameters.

55. If QR be a chord parallel to the tangent at P, and if QL, PN, RM be drawn parallel to one asymptote to meet the other,

$$CL \,.\, CM = CN^2.$$

56. If a circle touch the transverse axis at a focus, and pass through one end of the conjugate, the chord intercepted by the conjugate is a third proportional to the conjugate and transverse semi-axes.

57. A line through one of the vertices, terminated by two lines drawn through the other vertex parallel to the asymptotes, is bisected at the other point where it cuts the curve.

58. If PSQ be a focal chord, and if the tangents at P and Q meet in T, the difference between PTQ and half $PS'Q$ is a right angle.

59. If a straight line passing through a fixed point C meet two fixed lines OA, OB in A and B, and if P be taken in AB such that $CP^2 = CA \,.\, CB$, the locus of P is an hyperbola, having its asymptotes parallel to OA, OB.

60. If from the points P and Q in an hyperbola there be drawn PL, QM parallel to each other to meet one asymptote, and PR, QN also parallel to each other to meet the other asymptote, $PL \,.\, PR = QM \,.\, QN$.

61. Prove that the locus of the point of intersection of two tangents to a parabola which cut at a constant angle is an hyperbola, and that the angle between its asymptotes is double the external angle between the tangents.

62. An ordinate VQ of any diameter CP is produced to meet the asymptote in R, and the conjugate hyperbola in Q'; prove that

$$QV^2 + Q'V^2 = 2RV^2.$$

Prove also that the tangents at Q and Q' meet the diameter CP in points equidistant from C.

63. A chord QPL meets an asymptote in L, and a tangent from L is drawn touching at R; if PM, RE, QN, be drawn parallel to the asymptote to meet the other,

$$PM + QN = 2 \,.\, RE.$$

64. Tangents are drawn from any point in a circle through the foci; prove that the lines bisecting the angle between the tangents, or between one tangent and the other produced, all pass through a fixed point.

65. If a circle through the foci meet two confocal hyperbolas in P and Q, the angle between the tangents at P and Q is equal to PSQ.

66. If SY, $S'Y'$ be perpendiculars on the tangent at P, and if PN be the ordinate, the angles PNY, PNY' are supplementary.

67. Find the position of P when the area of the triangle YCY' is the greatest possible, and shew that, in that case,

$$PN \,.\, SC = BC^2.$$

68. If the tangent at P meet the conjugate axis in t, the areas of the triangles SPS', StS' are in the ratio of $CD^2 : St^2$.

69. If SY, SZ be perpendiculars on two tangents which meet in T, YZ is perpendicular to $S'T$.

70. A circle passing through a focus, and having its centre on the transverse axis, touches the curve; shew that the focal distance of the point of contact is equal to the latus rectum.

71. If CQ be conjugate to the normal at P, then is CP conjugate to the normal at Q.

72. From a point in the auxiliary circle lines are drawn touching the curve in P and P'; prove that SP, $S'P'$ are parallel.

73. If any hyperbola is drawn confocal with a given ellipse, and if PN is the ordinate of a point of intersection of the hyperbola with the ellipse, and NT the tangent from N to the auxiliary circle of the hyperbola, prove that the angle PNT is always the same.

74. Find the locus of the points of contact of tangents to a series of confocal hyperbolas from a fixed point in the axis.

75. Tangents to an hyperbola are drawn from any point in one of the branches of the conjugate, shew that the chord of contact will touch the other branch of the conjugate.

76. An ordinate NP meets the conjugate hyperbola in Q; prove that the normals at P and Q meet on the transverse axis.

77. A parabola and an hyperbola have a common focus S and their axes in the same direction. If a line SPQ cut the curves in P and Q, the angle between the tangents at P and Q is equal to half the angle between the axis and the other focal distance of the hyperbola.

78. If an hyperbola be described touching the four sides of a quadrilateral which is inscribed in a circle, and one focus lie on the circle, the other focus will also lie on the circle.

79. A conic section is drawn touching the asymptotes of an hyperbola. Prove that two of the chords of intersection of the curves are parallel to the chord of contact of the conic with the asymptotes.

80. A parabola P and an hyperbola H have a common focus, and the asymptotes of H are tangents to P; prove that the tangent at the vertex of P is a directrix of H, and that the tangent to P at the point of intersection passes through the further vertex of H.

81. From a given point in an hyperbola draw a straight line such that the segment intercepted between the other intersection with the hyperbola and a given asymptote shall be equal to a given line.

When does the problem become impossible?

82. If an ellipse and a confocal hyperbola intersect in P, an asymptote passes through the point on the auxiliary circle of the ellipse corresponding to P.

83. P is a point on an hyperbola whose foci are S and H; another hyperbola is described whose foci are S and P, and whose transverse axis is equal to $SP-2PH$: shew that the hyperbolas will meet only at one point, and that they will have the same tangent at that point.

84. A point D is taken on the axis of an hyperbola, of which the eccentricity is 2, such that its distance from the focus S is equal to the distance of S from the further vertex A'; P being any point on the curve, $A'P$ meets the latus rectum in K. Prove that DK and SP intersect on a certain fixed circle.

85. Shew that the locus of the point of intersection of tangents to a parabola, making with each other a constant angle equal to half a right angle, is an hyperbola.

86. The tangent and normal at any point intersect the asymptotes and axes respectively in four points which lie on a circle passing through the centre of the curve.

The radius of this circle varies inversely as the perpendicular from the centre on the tangent.

87. The difference between the sum of the squares of the distances of any point from the ends of any diameter and the sum of the squares of its distances from the ends of the conjugate is constant.

88. If a tangent meet the asymptotes in L and M, the angle subtended by LM at the farther focus is half the angle between the asymptotes.

89. If PN be the ordinate of P, and PT the tangent, prove that $SP : ST :: AN : AT$.

90. If an ellipse and an hyperbola are confocal, the asymptotes pass through the points on the auxiliary circle of the ellipse which correspond to the points of intersection of the two curves.

91. Two adjacent sides of a quadrilateral are given in magnitude and position; if the quadrilateral be such that a circle can be inscribed in it, the locus of the point of intersection of the other two sides is an hyperbola.

92. The tangent at P meets the conjugate axis in t, and tQ is perpendicular to SP; prove that SQ is of constant length.

93. An hyperbola, having a given transverse axis, has one focus fixed, and always touches a given straight line; the locus of the other focus is a circle.

94. A chord $PRVQ$ meets the directrices in R and V; shew that PR and VQ subtend, each at the focus nearer to it, angles of which the sum is equal to the angle between the tangents at P and Q.

95. A circle is drawn touching the transverse axis of an hyperbola at its centre, and also touching the curve; prove that the diameter conjugate to the diameter through either point of contact is equal to the distance between the foci.

96. A parabola is described touching the conjugate axes of an hyperbola at their extremities; prove that one asymptote is parallel to the axis of the parabola, and that the other asymptote is parallel to the chords of the parabola bisected by the first.

If a straight line parallel to the second asymptote meet the hyperbola and its conjugate in P, P', and the parabola in Q, Q', it may be shewn that $PQ = P'Q'$.

97. If two points E and E' be taken in the normal PG such that $PE = PE' = CD$, the loci of E and E' are hyperbolas having their axes equal to the sum and difference of the axes of the given hyperbola.

98. If two tangents are drawn to the same branch of an hyperbola, the external angle between them is half the difference between the angles which the chord of contact subtends at the foci.

If the tangents are drawn to opposite branches, the angle between them is half the sum, or half the difference, of these angles according as the points of contact are on the same or on opposite sides of the transverse axis.

99. Parabolas are drawn passing through two fixed points A and B, and having their axes in a given direction; find the locus of the foci, and, if a tangent be drawn at right angles to AB, prove that the locus of its point of contact P is an hyperbola.

100. Tangents are drawn from a point T to an hyperbola whose centre is C, and CT produced meets the hyperbola in P and the chord of contact of the tangents in V. If CD be the diameter conjugate to CP, and DT, DV meet the tangent at P in K and U, prove that the triangles PUV, TPK are equal in area.

101. One asymptote and three points P, Q, R of an hyperbola are given, construct the other asymptote.

102. If an ellipse be described having its centre on a given hyperbola, its foci on the asymptotes, and passing through the centre of the hyperbola, prove that the minor axis of the ellipse is equal to the major axis of the hyperbola, and the ellipse touches the minor axis of the hyperbola.

103. The angular point A of a triangle ABC is fixed, and the angle A is given, while the points B and C move on a fixed straight line; prove that the locus of the centre of the circle circumscribing the triangle is an hyperbola, and that the envelope of the circle is another circle.

104. Given an asymptote CQ and two points on an hyperbola, P, p on the curve, shew that the envelope of the axes is a parabola.

105. Find the locus of the middle points of a system of chords of an hyperbola, passing through a fixed point on one of the asymptotes.

106. If a conic be described having for its axes the tangent and normal at any point of a given ellipse, and touching at its centre the axis-major of the given ellipse, and if another conic be described in the same manner, but touching the minor axis at the centre, prove that the foci of these conics lie in two circles concentric with the given ellipse, and having their diameters equal to the sum and difference of its axes.

107. An ellipse and an hyperbola are confocal; if a tangent to one intersect at right angles a tangent to the other, the locus of the point of intersection is a circle.

Shew also that the difference of the squares on the distances from the centre of parallel tangents is constant.

108. If a circle passing through any point P of the curve, and having its centre on the normal at P, meets the curve again in Q and R, the tangents at Q and R intersect on a fixed straight line.

109. If the tangent at P meet an asymptote in T, the angle between that asymptote and $S'P$ is double the angle STP.

110. Four tangents to an hyperbola form a rectangle. If one side AB of the rectangle intersect a directrix in F, and S be the corresponding focus, the triangles FSA, FBS are similar.

111. An ellipse and hyperbola have the same transverse axis, and their eccentricities are the reciprocals of one another; prove that the tangents to each through the focus of the other intersect at right angles in two points and also meet the conjugate axis on the auxiliary circle.

112. ACA' and BCB' are the transverse and conjugate axes of an ellipse, of which S and S' are the foci. P is one of the points of intersection of this ellipse and a confocal hyperbola, and aCa' is the transverse axis of the hyperbola.

Prove that $SP = Aa$, $S'P = A'a$, and $aB = CP$.

113. Prove that if A, B and S are three given points, two parabolas can be drawn through A and B with S as focus, and that the axes of these parabolas are parallel to the asymptotes of the hyperbola which can be drawn through S with its foci at A and B.

CHAPTER V.

The Rectangular Hyperbola.

If the axes of an hyperbola be equal, the angle between the asymptotes is a right angle, and the curve is called equilateral or rectangular.

135. Prop. I. *In a rectangular hyperbola*

$$CS^2 = 2AC^2, \text{ and } SA^2 = 2AX^2.$$

For
$$CS^2 = AC^2 + BC^2 = 2AC^2,$$
and
$$SA : AX :: SC : AC;$$
$$\therefore SA^2 = 2AX^2.$$

Observe that, in the figure of Art. 102, SDC is an isosceles triangle, since

$$SD = BC, \text{ and } CD = AC,$$

and therefore
$$SD = DC.$$

136. Prop. II. *The asymptotes of a rectangular hyperbola bisect the angles between any pair of conjugate diameters.*

For, in a rectangular or equilateral hyperbola,

$$CA = CB,$$

and therefore, since
$$CP^2 - CD^2 = CA^2 - CB^2,$$
$$CP = CD,$$

CP, CD being any conjugate semi-diameters.

Also, in the figure of Art. 123, the parallelogram $CPLD$ is a rhombus, and therefore CL bisects the angle PCD.

Cor. Supplemental chords are equally inclined to the asymptotes, for they are parallel to conjugate diameters.

137. Prop. III. *If CY be the perpendicular from the centre on the tangent at P, the angle PCY is bisected by the transverse axis, and half the transverse axis is a mean proportional between CY and CP.*

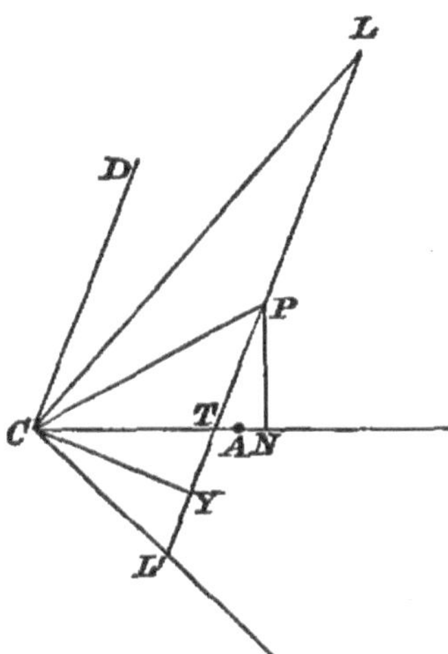

For the angle $PCL = DCL$

$$= YCL';$$

$$\therefore PCA = ACY.$$

Hence it follows that the triangles PCN, TCY are similar, and that

$$CY : CT :: CN : CP;$$

$$\therefore CY \,.\, CP = CT \,.\, CN = AC^2.$$

Hence also, if we join PA and AY, we observe that the triangles PAC, AYC are similar.

138. Prop. IV. *Diameters at right angles to each other are equal.*

Let CP, CP' be semi-diameters at right angles to each other, and CD conjugate to CP.

Then, if CL, CL' be the asymptotes,

$$\text{the angle } P'CL' = PCL = DCL;$$

$$\therefore CP' = CD = CP.$$

Hence it follows, by help of the theorem of Art. 120, that focal chords at right angles to each other are equal, and that focal chords parallel to conjugate diameters are equal.

139. Prop. V. *If the normal at P meet the axes in G and g,*

$$CN = NG \text{ and } PG = Pg = CD,$$

CD being conjugate to CP.

For (Art. 115) $NG : CN :: BC^2 : AC^2$;

$$\therefore NG = CN.$$

Also $\quad PF \,.\, PG = BC^2$ and $PF \,.\, Pg = AC^2$;

$$\therefore PG = Pg.$$

Further (Art. 128) $PG : CD :: BC : AC$;

$$\therefore PG = CD = CP.$$

140. PROP. VI. *If QV be an ordinate of a diameter PCp,*

$$QV^2 = PV \,.\, Vp.$$

For
$$QV^2 : PV \,.\, Vp :: CD^2 : CP^2,$$
and
$$CD = CP;$$
$$QV^2 = PV \,.\, Vp = CV^2 - CP^2.$$

141. PROP. VII. *The angle between a chord PQ, and the tangent at P, is equal to the angle subtended by PQ at the other extremity of the diameter through P.*

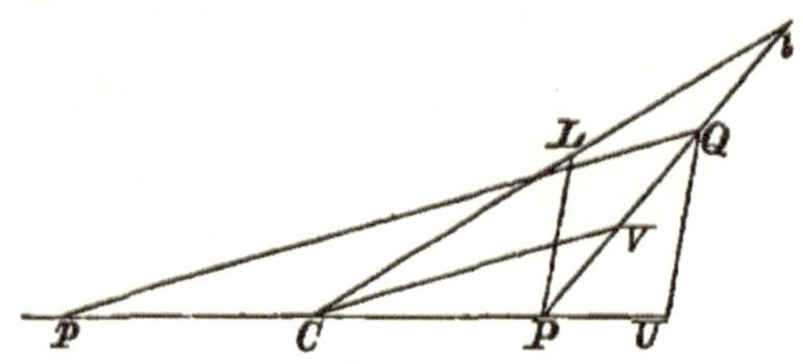

Let PQ and the tangent at P meet the asymptote in l and L. Then, if CV be conjugate to PQ,

$$\text{the angle } LPQ = PLC - VlC = LCP - VCl$$
$$= VCP = QpP.$$

Or thus, let QU parallel to the tangent at P, meet CP produced in U. Then

$$QU^2 = PU \,.\, Up,$$

or,

$$QU : PU :: Up : UQ.$$

Therefore the triangles PQU, QUp are similar, and the angle $QpU = PQU = LPQ$.

If P and Q are on opposite branches of the curve, the same proof shews that

$$\text{the angle } QpU = UQP = LPQ;$$

$$\therefore QPL' = QpP.$$

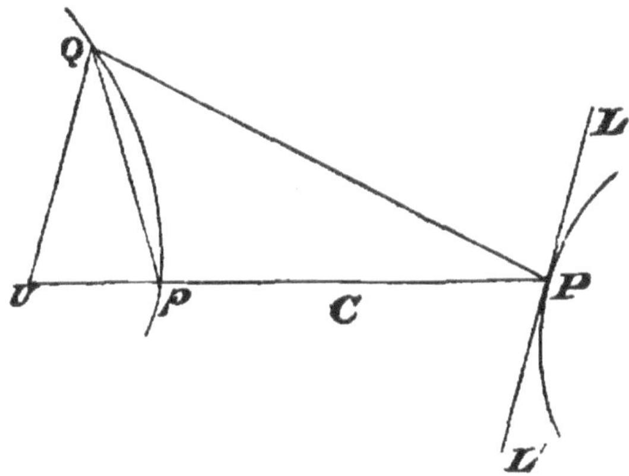

If QP is the normal at P, it follows that QP subtends a right angle at the other end of the diameter through P.

142. PROP. VIII. *Any chord subtends, at the ends of any diameter, angles which are equal or supplementary.*

This theorem divides itself into four cases, which are shewn in the appended figures.

Let QR be the chord, and Pp the diameter. Then, if LP be the tangent at P, fig. (1),

$$\text{the angle } LPQ = QpP,$$
$$\text{and } LPR = RpP;$$
$$\therefore QPR = QpR.$$

In fig. (2), if pl be the tangent at p, parallel to PL,

$$QpR = Qpl + lpR = Qpl + pPR,$$

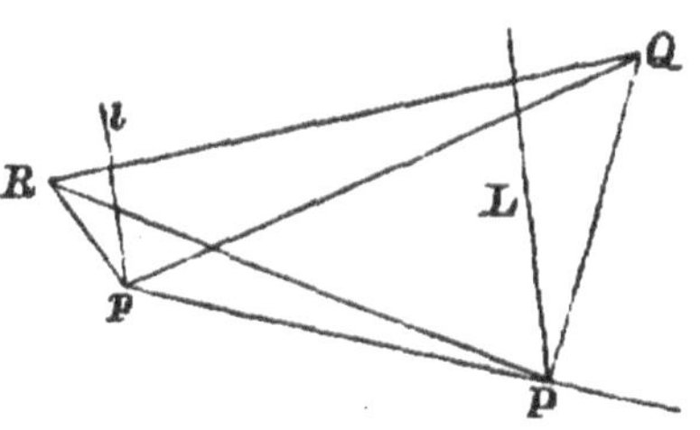

and $$QPR = QPL + LPR = QpP + LPR;$$
$$\therefore QpR + QPR = lpP + LPp,$$
that is, QpR and QPR are together equal to two right angles.

In fig. (3)
$$\begin{aligned} QPR &= QPL + LPp + pPR \\ &= QpP + Ppl + lpR \\ &= QpR. \end{aligned}$$

In fig. (4) $QPL = QpP$, and $RPL' = RpP$;
$$\therefore QpR = QPL + RPL';$$
therefore QpR and QPR are together equal to two right angles.

Hence it will be seen that when QR, or QR produced, meet the diameter Pp between P and p, the angles subtended at P and p are equal; in other cases they are supplementary.

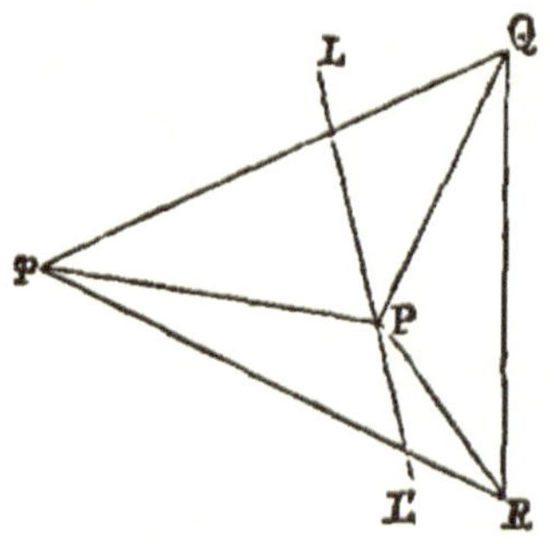

In the cases of the second and third figures, if one of the angles QPR is a right angle, the other angle QpR is also a right angle. The four points Q, P, p, R are then concyclic, and QR is a diameter of the circle.

143. Prop. IX. *If a rectangular hyperbola circumscribe a triangle, it passes through the orthocentre.*

Note. *The orthocentre is the point of intersection of the perpendiculars from the angular points on the opposite sides.*

If O be the orthocentre, the triangles LOP, LQR are similar, and
$$LO : LP :: LQ : LR;$$
$$\therefore LO \,.\, LR = LP \,.\, LQ.$$

But, if a rectangular hyperbola pass through P, Q, R, the diameters parallel to LR, PQ are equal: hence O is a point on the curve.

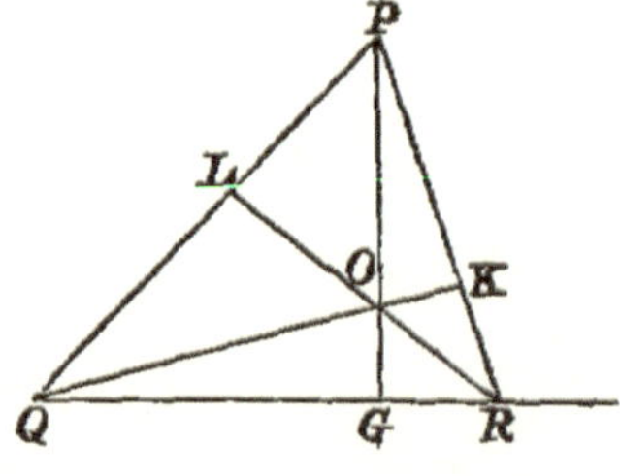

If the angle PRQ is a right angle, the line ROL will be the tangent to the curve at R, so that if a rectangular hyperbola pass through the angular points of a right-angled triangle, the hypotenuse will be parallel to the normal at the right-angle vertex.

144. Prop. X. *If a rectangular hyperbola circumscribe a triangle, the locus of its centre is the nine-point circle of the triangle.*

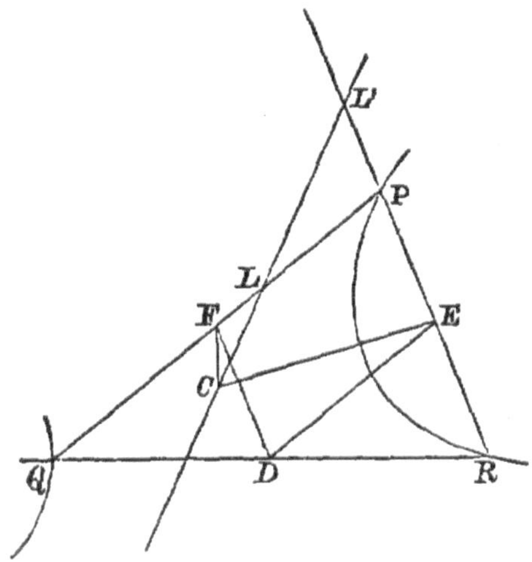

If PQR be the triangle, let L, L' be the points in which an asymptote meets the sides PQ, PR.

Join C, the centre of the hyperbola, with E and F, the middle points of PR and PQ.

Then CF is conjugate to PQ, and CE to PR; therefore the angle

$$\begin{aligned} FCE &= FCL + L'CE = CLF + EL'C \\ &= PLL' + PL'L = FPE \\ &= FDE, \end{aligned}$$

if D be the middle point of QR.

$\therefore D$, E, F, C are concyclic; that is, C lies on the nine-point circle.

A similar proof is applicable to the case in which the points P, Q, R lie on the same branch of the hyperbola.

EXAMPLES.

1. PCP is a transverse diameter, and QV an ordinate; shew that QV is the tangent at Q to the circle circumscribing the triangle PQp.

2. If the tangent at P meet the asymptotes in L and M, and the normal meet the transverse axis in G, a circle can be drawn through C, L, M, and G, and LGM is a right angle.

3. If AA' be any diameter of a circle, PP' any ordinate to it, then the locus of the intersections of AP, $A'P'$ is a rectangular hyperbola.

4. Given an asymptote and a tangent at a given point, construct the rectangular hyperbola.

5. The points of intersection of an ellipse and a confocal rectangular hyperbola are the extremities of the equi-conjugate diameters of the ellipse.

6. If CP, CD be conjugate semi-diameters, and PN, DM ordinates of any diameter, the triangles PCN, DCM are equal in all respects.

7. The distance of any point from the centre is a geometric mean between its distances from the foci.

8. If P be a point on an equilateral hyperbola, and if the tangent at Q meet CP in T, the circle circumscribing CTQ touches the ordinate QV conjugate to CP.

9. If a circle be described on SS' as diameter, the tangents at the vertices will intersect the asymptotes in the circumference.

10. If two concentric rectangular hyperbolas be described, the axes of one being the asymptotes of the other, they will intersect at right angles.

11. If the tangents at two points Q and Q' meet in T, and if CQ, CQ' meet these tangents in R and R', the points R, T, R', C are concyclic.

12. If from a point Q in the conjugate axis QA be drawn to the vertex, and QR parallel to the transverse axis to meet the curve, $QR = AQ$.

13. Straight lines, passing through a given point, are bounded by two fixed lines at right angles to each other; find the locus of their middle points.

14. Given a point Q and a straight line AB, if a line QCP be drawn cutting AB in C, and P be taken in it, so that, PD being a perpendicular upon AB, CD may be of constant magnitude, the locus of P is a rectangular hyperbola.

15. Every conic passing through the centres of the four circles which touch the sides of a triangle, is a rectangular hyperbola.

16. Ellipses are inscribed in a given parallelogram, shew that their foci lie on a rectangular hyperbola.

17. If two focal chords be parallel to conjugate diameters, the lines joining their extremities intersect on the asymptotes.

18. If P, Q be two points of a rectangular hyperbola, centre O, and QN the perpendicular let fall on the tangent at P, the circle through O, N, and P will pass through the middle point of the chord P, Q.

19. Having given the centre, a tangent, and a point of a rectangular hyperbola, construct the asymptotes.

20. If a right-angled triangle be inscribed in the curve, the normal at the right angle is parallel to the hypotenuse.

21. On opposite sides of any chord of a rectangular hyperbola are described equal segments of circles; shew that the four points, in which the circles, to which these segments belong, again meet the hyperbola, are the angular points of a parallelogram.

22. Two lines of given lengths coincide with and move along two fixed lines, in such a manner that a circle can always be drawn through their extremities; the locus of the centre is a rectangular hyperbola.

23. If a rectangular hyperbola, having its asymptotes coincident with the axes of an ellipse, touch the ellipse, the axis of the hyperbola is a mean proportional between the axes of the ellipse.

24. The tangent at a point P of a rectangular hyperbola meets a diameter QCQ' in T. Shew that CQ and TQ' subtend equal angles at P.

25. If A be any point in a rectangular hyperbola, of which O is the centre, BOC the straight line through O at right angles to OA, D any other point in the curve, and DB, DC parallel to the asymptotes, prove that B, D, A, C are concyclic.

26. The angle subtended by any chord at the centre is the supplement of the angle between the tangents at the ends of the chord.

27. If two rectangular hyperbolas intersect in A, B, C, D; the circles described on AB, CD as diameters intersect each other orthogonally.

28. Prove that the triangle, formed by the tangent at any point and its intercepts on the axes, is similar to the triangle formed by the straight line joining that point with the centre, and the abscissa and ordinate of the point.

29. The angle of inclination of two tangents to a parabola is half a right angle; prove that the locus of their point of intersection is a rectangular hyperbola, having one focus and the corresponding directrix coincident with the focus and directrix of the parabola.

30. P is a point on the curve, and PM, PN are straight lines making equal angles with one of the asymptotes; if MP, NP be produced to meet the curve in P' and Q', then $P'Q'$ passes through the centre.

31. A circle and a rectangular hyperbola intersect in four points and one of their common chords is a diameter of the hyperbola; shew that the other common chord is a diameter of the circle.

32. AB is a chord of a circle and a diameter of a rectangular hyperbola; P any point on the circle; AP, BP, produced if necessary, meet the hyperbola in Q, Q', respectively; the point of intersection of BQ, AQ' will be on the circle.

33. PP' is any diameter, Q any point on the curve, PR, $P'R'$ are drawn at right angles to PQ, $P'Q$ respectively, intersecting the normal at Q in R, R'; prove that QR and QR' are equal.

34. Parallel tangents are drawn to a series of confocal ellipses; prove that the locus of the points of contact is a rectangular hyperbola having one of its asymptotes parallel to the tangents.

35. If tangents, parallel to a given direction, are drawn to a system of circles passing through two fixed points, the points of contact lie on a rectangular hyperbola.

36. If from a point P on the curve chords are equally inclined to the asymptotes, the line joining their other extremities passes through the centre.

37. From the point of intersection of the directrix with one of the asymptotes of a rectangular hyperbola a tangent is drawn to the curve and meets the other asymptote in T; shew that CT is equal to the transverse axis.

38. The normals at the ends of two conjugate diameters intersect on the asymptote, and are parallel to another pair of conjugate diameters.

39. If the base AB of a triangle ABC be fixed, and if the difference of the angles at the base is constant, the locus of the vertex is a rectangular hyperbola.

40. A circle described through the angular points A, B of a given triangle ABC meets AC in D. If BD meet the tangent at A in P, shew that the vertex and orthocentre of the triangle APB lie on fixed rectangular hyperbolas.

41. The locus of the point of intersection of tangents to an ellipse which make equal angles with the transverse and conjugate axes respectively, and are not at right angles, is a rectangular hyperbola whose vertices are the foci of the ellipse.

42. If OT is the tangent at the point O of a rectangular hyperbola, and PQ a chord meeting it at right angles in T, the two bisectors of the angle OCT bisect OP and OQ.

43. With two sides of a square as asymptotes, and the opposite point as focus, a rectangular hyperbola is described; prove that it bisects the other sides.

44. With the focus S of a rectangular hyperbola as centre and radius equal to SC a circle is described, prove that it touches the conjugate hyperbola.

45. If parallel normal chords are drawn to a rectangular hyperbola, the diameter bisecting them is perpendicular to the join of their feet.

46. From the foot of the ordinate PN of a point P of a rectangular hyperbola, tangents NQ, NR are drawn to the circle on AA' as diameter. Prove that PQ passes through A', and PR through A, and that, if QR intersect AA' in M, PM is the tangent at P.

47. Shew that the angle between two tangents to a rectangular hyperbola is equal or supplementary to the angle which their chord of contact subtends at the centre, and that the bisectors of these angles meet on the chord of contact.

48. Through a point P on an equilateral hyperbola two lines are drawn parallel to a pair of conjugate diameters; the one meeting the curve in P, P', and the other meeting the asymptotes in Q, Q'; shew that $PP' = QQ'$.

49. If four points forming a parallelogram be taken on a rectangular hyperbola, then the product of the perpendiculars from any point of the curve on one pair of opposite sides equals the product of the perpendiculars on the other pair of sides.

CHAPTER VI.

THE CYLINDER AND THE CONE.

DEFINITION.

145. If a straight line move so as to pass through the circumference of a given circle, and to be perpendicular to the plane of the circle, it traces out a surface called a *Right Circular Cylinder.* The straight line drawn through the centre of the circle perpendicular to its plane is the *Axis* of the Cylinder.

It is evident that a section of the surface by a plane perpendicular to the axis is a circle, and that a section by any plane parallel to the axis consists of two parallel lines.

PROP. I. *Any section of a cylinder by a plane not parallel or perpendicular to the axis is an ellipse.*

If APA' be the section, let the plane of the paper be the plane through the axis perpendicular to APA'.

Inscribe in the cylinder a sphere touching the cylinder in the circle EF and the plane APA' in the point S.

Let the planes APA', EF intersect in XK, and from any point P of the section draw PK perpendicular to XK.

Draw through P the circular section QP, cutting APA' in PN, so that PN is at right angles to AA' and therefore parallel to XK.

Let the generating line through P meet the circle EF in R; and join SP.

Then PS and PR are tangents to the sphere;

$$\therefore SP = PR = EQ.$$

But $$EQ : NX :: AE : AX$$
$$:: SA : AX,$$
and $$NX = PK,$$
$$\therefore SP : PK :: SA : AX.$$

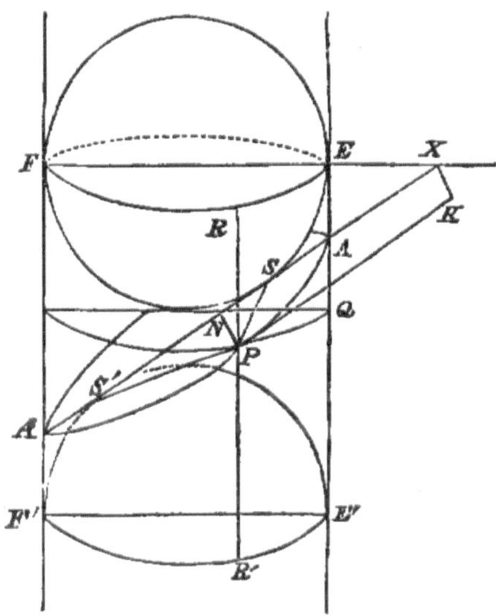

Also, AE being less than AX, SA is less than AX, and the curve APA' is therefore an ellipse, of which S is the focus and XK the directrix.

If another sphere be inscribed in the cylinder touching AA' in S', S' is the other focus, and the corresponding directrix is the intersection of the plane of contact $E'F'$ with APA'.

Producing the generating line RP to meet the circle $E'F'$ in R' we observe that $S'P = PR'$, and therefore

$$\begin{aligned} SP + S'P &= RR' = EE' \\ &= AE + AE' \\ &= AS + AS'; \end{aligned}$$

and $$AS' = AE' = A'F = A'S,$$
$$\therefore SP + S'P = AA'.$$

The transverse axis of the section is AA' and the conjugate, or minor, axis is evidently a diameter of a circular section.

146. DEF. If O be a fixed point in a straight line OE drawn through the centre E of a fixed circle at right angles to the plane of the circle, and if a straight line QOP move so as always to pass through the circumference of the circle, the surface generated by the line QOP is called a *Right Circular Cone.*

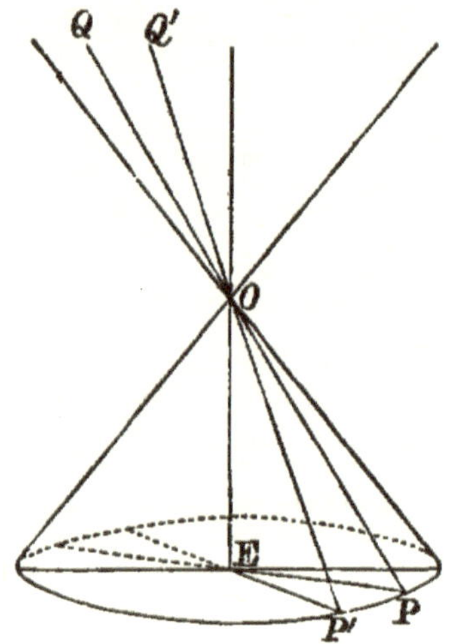

The line OE is called the axis of the cone, the point O is the *vertex*, and the constant angle POE is the semi-vertical angle of the cone.

It is evident that any section by a plane perpendicular to the axis, or parallel to the base of the cone, is a circle; and that any section by a plane through the vertex consists of two straight lines, the angle between which is greatest and equal to the vertical angle when the plane contains the axis.

Any plane containing the axis is called a *Principal Section.*

147. PROP. II. *The section of a cone by a plane, which is not perpendicular to the axis, and does not pass through the vertex, is either an Ellipse, a Parabola, or an Hyperbola.*

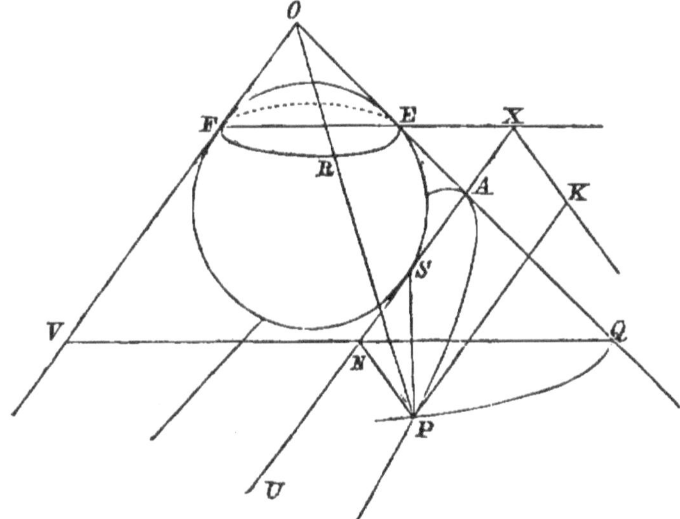

Let UAP be the cutting plane, and let the plane of the paper be that principal section which is perpendicular to the plane UAP; OV, OAQ being the generating lines in the plane of the paper.

Let AU be the intersection of the principal section VOQ by the plane PAU perpendicular to it, and cutting the cone in the curve AP.

Inscribe a sphere in the cone, touching the cone in the circle EF and the plane AP in the point S, and let XK be the intersection of the planes AP, EF. Then XK is perpendicular to the plane of the paper.

Taking any point P in the curve, join OP cutting the circle EF in R, and join SP.

Draw through P the circular section QPV cutting the plane AP in PN which is therefore perpendicular to AN and parallel to XK.

Then, SP and PR being tangents to the sphere,

$$SP = PR = EQ;$$

and

$$EQ : NX :: AE : AX$$
$$:: AS : AX.$$

Also

$$NX = PK;$$

$$\therefore SP : PK :: SA : AX.$$

The curve AP is therefore an Ellipse, Parabola, or Hyperbola, according as SA is less than, equal to, or greater than AX. In any case the point S is a focus and the corresponding directrix is the intersection of the plane of the curve with the plane of contact of the sphere.

148. (1) If AU be parallel to OV, the angle

$$AXE = OFE = OEF = AEX,$$

so that

$$SA = AE = AX;$$

the section is therefore a parabola when the cutting plane is parallel to a generating line, and perpendicular to the principal section which contains the generating line.

(2) Let the line AU meet the curve again in the point A' on the same side of the vertex as the point A.

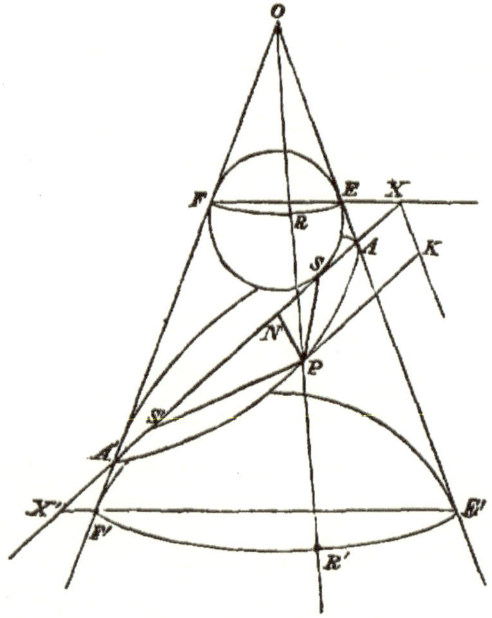

Then the angle

$$AEX = OFX$$
$$> FXA,$$

and therefore $\qquad AE < AX,$

that is $\qquad SA < AX,$

and the curve is an ellipse.

In this case another sphere can be inscribed in the cone, touching the cone along the circle $E'F'$ and touching the plane AP in S'.

It may be shewn as before that S' is a focus and that the corresponding directrix is the intersection of the planes $E'F'$, APA'.

(3) Let the line UA produced meet the cone on the other side of the vertex. The section then consists of two separate branches.

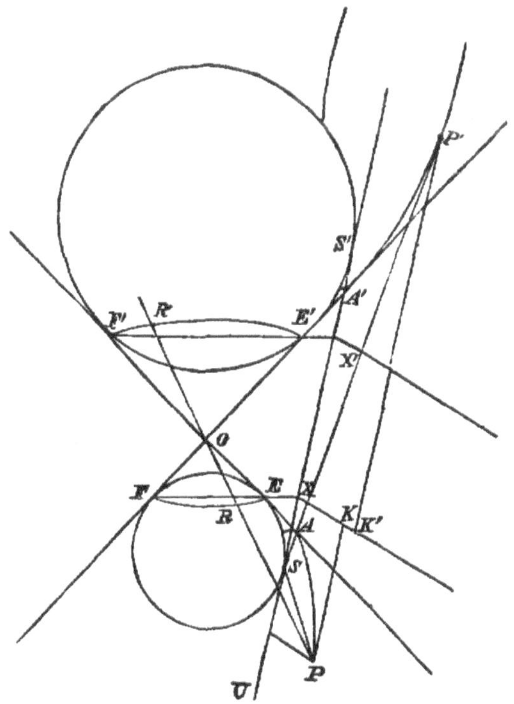

Also the angle $AEX = A'FX$

$< AXF,$

and therefore $AE > AX,$

that is $AS > AX,$

and the curve AP is one branch of an hyperbola, the other branch being the section $A'P'$.

Taking P' in the other branch the proof is the same as before that

$$SP' : P'K' :: SA : AX.$$

In this case a sphere can be inscribed in the other branch of the cone, touching the cone along the circle $E'F'$, and the plane $UA'P'$ in S', and it can be shewn that S' is the other focus of the hyperbola, and that the directrix is the intersection of the cutting plane with the plane of contact $E'F'$.

Hence the section of a cone by a plane cutting in AU the principal section VOQ perpendicular to it is an Ellipse, Parabola, or Hyperbola, according as the angle EAX is greater than, equal to, or less than, the vertical angle of the cone.

Further, it is obvious that, if any plane be drawn parallel to the plane AP, the ratio of AE to AX is always the same; hence it follows that all parallel sections have the same eccentricity.

149. This method of determining the focus and directrix was published by Mr Pierce Morton, of Trinity College, in the first volume of the Cambridge *Philosophical Transactions*.

The method was very nearly obtained by Hamilton, who gave the following construction.

First finding the vertex and focus, A and S, take AE along the generating line equal to AS, and draw the circular section through E; the directrix will be the line of intersection of the plane of the circle with the given plane of section.

Hamilton also demonstrated the equality of SP and PR.

150. PROP. III. *To prove that, in the case of an elliptic section,*

$$SP + S'P = AA'.$$

Taking the 2nd figure,

$$SP = PR \text{ and } S'P = PR';$$
$$\therefore SP + S'P = RR' = EE'$$
$$= AE + AE'$$
$$= AS + AS'.$$

But $$A'S' = A'F' = FF' - A'F$$
$$= EE' - A'S,$$

also $$A'S' + SS' = A'S;$$
$$\therefore 2A'S' + SS' = EE'.$$

Similarly $$2AS + SS' = EE';$$
$$\therefore A'S' = AS,$$

and $$AS' = A'S.$$

Hence $$SP + S'P = AA',$$

and the transverse, or major axis $= EE'$.

In a similar manner it can be shewn that in an hyperbolic section

$$S'P - SP = AA'.$$

151. Prop. IV. *To shew that, in a parabolic section,*

$$PN^2 = 4AS \,.\, AN.$$

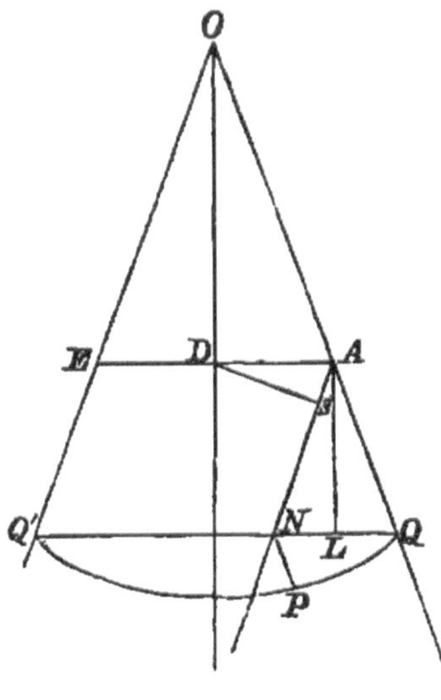

Let A be the vertex of the section, and let ADE be the diameter of the circular section through A. From D let fall DS perpendicular to AN;

then
$$\begin{aligned} PN^2 &= QN \,.\, NQ' \\ &= QN \,.\, AE \\ &= 4NL \,.\, AD, \end{aligned}$$

if AL be perpendicular to NQ.

But the triangles ANL, ADS being similar,
$$NL : AN :: AS : AD;$$
$$\therefore NL \,.\, AD = AN \,.\, AS,$$

and
$$PN^2 = 4AS \,.\, AN.$$

152. Prop. V. *To shew that, in an elliptic section, PN^2 is to $AN \,.\, NA'$ in a constant ratio.*

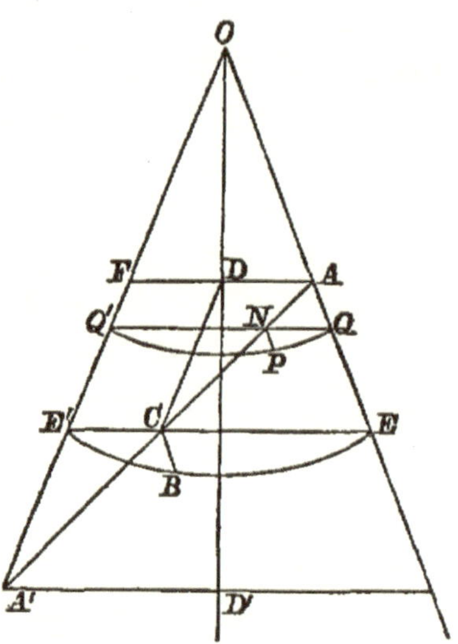

Draw through P the circular section QPQ', bisect AA' in C, and draw through C the circular section EBE'.

Then
$$QN : AN :: CE : AC,$$

and
$$NQ' : NA' :: CE' : A'C;$$
$$\therefore QN \,.\, NQ' : AN \,.\, NA' :: EC \,.\, CE' : AC^2,$$

or
$$PN^2 : AN \,.\, NA' :: EC \,.\, CE' : AC^2;$$

and, the transverse axis being AA', the square of the semi-minor axis $= BC^2 = EC.CE'$. Again, if ADF be perpendicular to the axis, $AD = DF$, and, AC being equal to CA', CD is parallel to $A'F$, and therefore
$$CE' = FD = AD.$$

Similarly, $CE = A'D'$, the perpendicular from A' on the axis;
$$\therefore BC^2 = AD \,.\, A'D',$$

that is, *the semi-minor axis is a mean proportional between the perpendiculars from the vertices on the axis of the cone.*

COR. If H, H' are the centres of the focal spheres, the angles HAH', $HA'H'$ are right angles, so that H, A, H', A' are concyclic.

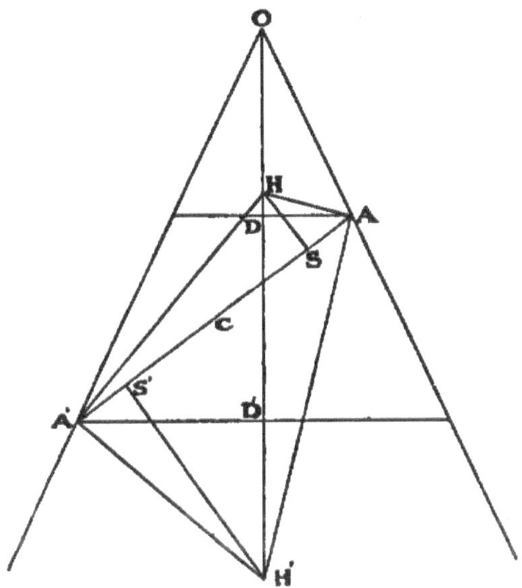

It follows that the triangles ASH, $A'H'D'$ are similar, as are also the triangles $A'S'H'$, AHD, so that

$$SH : A'D' :: AH : A'H' :: AD : S'H';$$

and
$$SH \,.\, S'H' = AD \,.\, A'D' = BC^2;$$

$\therefore$ *the semi-minor axis is a mean proportional between the radii of the focal spheres.*

The fact that H, A, H', A' are concyclic also shews that the sphere of which HH' is a diameter intersects the plane of the ellipse in its auxiliary circle.

153. In exactly the same manner it can be shewn that, for an hyperbolic section,

$$PN^2 : AN \,.\, NA' :: CE \,.\, CE' : AC^2,$$

and that
$$CE = AD, \text{and } CE' = A'D'.$$

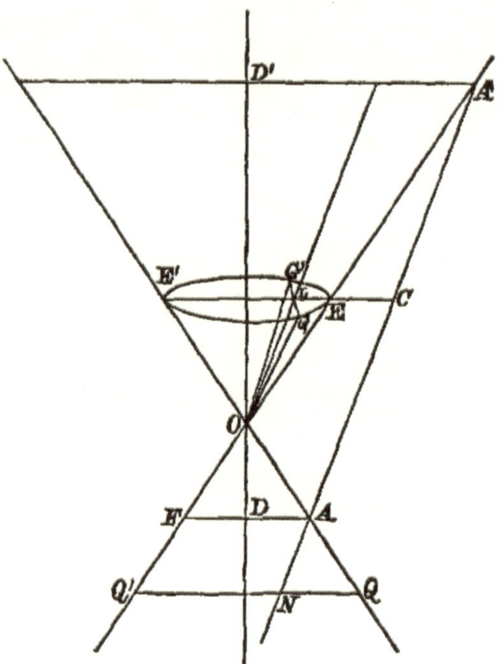

Also, as in the case of the ellipse, BC is a mean proportional between AD and $A'D'$, and is also a mean proportional between the radii of the focal spheres.

154. Prop. VI. *The two straight lines in which a cone is intersected by a plane through the vertex parallel to an hyperbolic section are parallel to the asymptotes of the hyperbola.*

Taking the preceding figure, let the parallel plane cut the cone in the lines OG, OG', and the circular section through C in the line GLG', which will be perpendicular to the plane of the paper, and therefore perpendicular to EE' and to OL.

Hence $$GL^2 = EL \,.\, E'L.$$

But $$EL : EC :: OL : A'C,$$
and $$E'L : E'C :: OL : AC;$$
$$\therefore GL^2 : EC \,.\, E'C :: OL^2 : AC^2,$$
or $$GL : OL :: BC : AC;$$
therefore, (Art. 102), OG and OG' are parallel to the asymptotes of the hyperbola.

Hence, for all parallel hyperbolic sections, the asymptotes are parallel to each other.

If the hyperbola be rectangular, the angle GOG' is a right angle; but this is evidently not possible if the vertical angle of the cone be less than a right angle.

When the vertical angle of the cone is not less than a right angle, and when GOG' is a right angle, LOG is half a right angle, and therefore
$$OL = LG,$$
and $$2 \,.\, OL^2 = OG^2 = OE^2,$$
and the length OL is easily constructed.

Hence, placing OL, and drawing the plane GOG' perpendicular to the principal section through OL, any section by a plane parallel to GOG' is a rectangular hyperbola.

It will be observed that the eccentricity of the section is greatest when its plane is parallel to the axis of the cone.

155. PROP. VII. *The sphere which passes through the circles of contact of the focal spheres with the surface of the cone intersects the plane of the section in its director circle.*

Let Q, Q' be the points in which the straight line AA' is intersected by the sphere which passes through the circles EF and $E'F'$.

Then the sphere intersects the plane of the ellipse in the circle of which QQ' is the diameter.

Also $$CQ^2 - CA^2 = AQ \,.\, AQ' = AE \,.\, AE'$$
$$= AS \,.\, AS' = BC^2;$$
$$\therefore CQ^2 = AC^2 + BC^2,$$
so that CQ is the radius of the director circle.

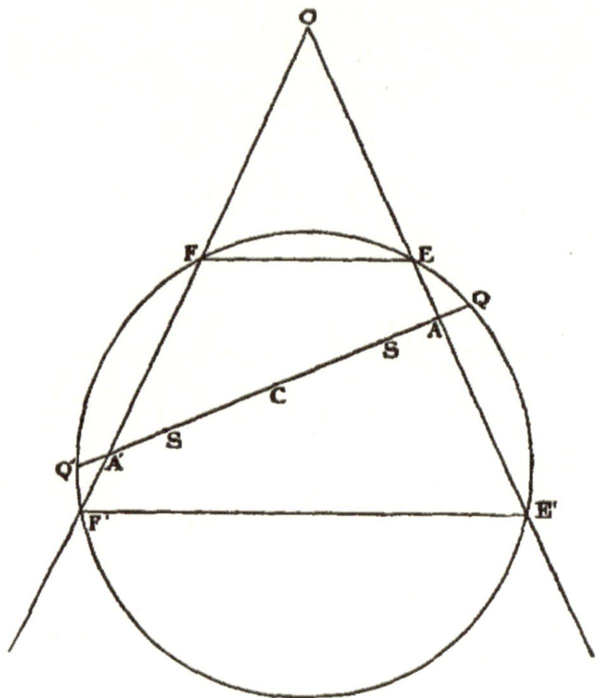

Changing the figure the proof is exactly the same for the hyperbola.

156. PROP. VIII. *If two straight lines be drawn through any point, parallel to two fixed lines, and intersecting a given cone, the ratio of the rectangles formed by the segments of the lines will be independent of the position of the point.*

Thus, if through E, the lines EPQ, $EP'Q'$ be drawn, parallel to two given lines, and cutting the cone in the points P, Q and P', Q', the ratio of $EP . EQ$ to $EP' . EQ'$ is constant.

Through O draw OK parallel to the given line to which EPQ is parallel, and let the plane through OK, EPQ, which contains the generating lines OP, OQ, meet the circular section through E in R and S, and the plane base in the straight line DFK, cutting the circular base in D and F.

Then DFK and ERS being sections of parallel planes by a plane are parallel to each other.

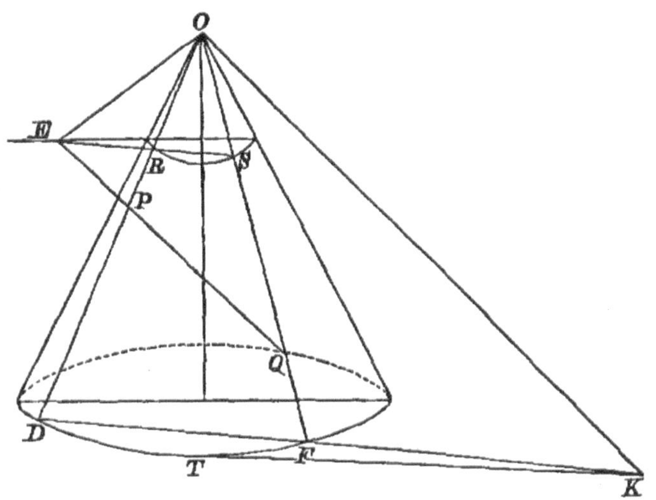

Also, EPQ is parallel to OK;

Therefore ERP, ODK are similar triangles, as are also ESQ, OFK;

$$\therefore EP : ER :: OK : DK,$$

and
$$EQ : ES :: OK : FK;$$

$$\therefore EP \,.\, EQ : ER \,.\, ES :: OK^2 : DK \,.\, FK$$
$$:: OK^2 : KT^2,$$

if KT be the tangent to the circular base from K.

If a similar construction be made for $EP'Q'$, we shall have

$$EP' \,.\, EQ' : ER' \,.\, ES' :: OK'^2 : K'T'^2.$$

But
$$ER \,.\, ES = ER' \,.\, ES';$$

therefore the rectangles $EP \,.\, EQ$ and $EP' \,.\, EQ'$ are each in a constant ratio to the same rectangle, and are therefore in a constant ratio to each other.

Since the plane through EPQ, $EP'Q'$ cuts the cone in an ellipse, parabola, or hyperbola, this theorem includes as particular cases those of Arts. 51, 58, 82, 92, 96, 124 and 134.

The proof is the same if the point P be within the cone, or if one or both of the lines meet opposite branches of the cone.

If the chords be drawn through the centre of the section PEP', the rectangles become the squares of the semi-diameters.

Hence the parallel diameters of all parallel sections of a cone are proportional to each other.

If the lines move until they become tangents the rectangles then become the squares of the tangents; therefore if a series of points be so taken that the tangents from them are parallel to given lines, these tangents are always in the same proportion. The locus of the point E will be the line of intersection of two fixed planes touching the cone, that is, a fixed line through the vertex.

EXAMPLES.

1. Shew how to cut from a cylinder an ellipse of given eccentricity.

2. What is the locus of the foci of all sections of a cylinder of a given eccentricity?

3. Shew how to cut from a cone an ellipse of given eccentricity.

4. Prove that all sections of a cone by parallel planes are conics of the same eccentricity.

5. What is the locus of the foci of the sections made by planes inclined to the axis at the same angle?

6. Find the least angle of a cone from which it is possible to cut an hyperbola, whose eccentricity shall be the ratio of two to one.

7. The centre of a spherical ball is moveable in a vertical plane which is equidistant from two candles of the same height on a table; find its locus when the two shadows on the ceiling are always just in contact.

8. Through a given point draw a plane cutting a given cone in a section which has the given point for a focus.

9. If the vertical angle of a cone, vertex O, be a right angle, P any point of a parabolic section, and PN perpendicular to the axis of the parabola,

$$OP = 2AS + AN,$$

A being the vertex and S the focus.

10. Prove that the directrices of all parabolic sections of a cone lie in the tangent planes of a cone having the same axis.

11. If the curve formed by the intersection of any plane with a cone be projected upon a plane perpendicular to the axis; prove that the curve of projection will be a conic section having its focus at the point in which the axis meets the plane of projection.

12. Prove that the latera recta of parabolic sections of a right circular cone are proportional to the distances of their vertices from the vertex of the cone.

13. The shadow of a ball is cast by a candle on an inclined plane in contact with the ball; prove that as the candle burns down, the locus of the centre of the shadow will be a straight line.

14. The vertex of a right cone which contains a given ellipse lies on a certain hyperbola, and the axis of the cone will be a tangent to the hyperbola.

15. Find the locus of the vertices of the right circular cones which can be drawn so as to pass through a given fixed hyperbola, and prove that the axis of the cone is always tangential to the locus.

16. An ellipse and an hyperbola are so situated that the vertices of each curve are the foci of the other, and the curves are in planes at right angles to each other. If P be a point on the ellipse, and O a point on the hyperbola, S the vertex, and A the interior focus of that branch of the hyperbola, then

$$AS + OP = AO + SP.$$

17. The latus rectum of any plane section of a given cone is proportional to the perpendicular from the vertex on the plane.

18. If a sphere is described about the vertex of a right cone as centre, the latera recta of all sections made by tangent planes to the sphere are equal.

19. Different elliptic sections of a right cone are taken such that their minor axes are equal; shew that the locus of their centres is the surface formed by the revolution of an hyperbola about the axis of the cone.

20. If two cones be described touching the same two spheres, the eccentricities of the two sections of them made by the same plane bear to one another a ratio constant for all positions of the plane.

21. If elliptic sections of a cone be made such that the volume between the vertex and the section is always the same, the minor axis will be always of the same length.

22. The vertex of a cone and the centre of a sphere inscribed within it are given in position: a plane section of the cone, at right angles to any generating line of the cone, touches the sphere: prove that the locus of the point of contact is a surface generated by the revolution of a circle, which touches the axis of the cone at the centre of the sphere.

23. Given a right cone and a point within it, there are two sections which have this point for focus; and the planes of these sections make equal angles with the straight line joining the given point and the vertex of the cone.

24. Prove that the centres of all plane sections of a cone, for which the distance between the foci is the same, lie on the surface of a right circular cylinder.

CHAPTER VII.

THE SIMILARITY OF CONICS, THE AREAS OF CONICS, AND THE CURVATURES OF CONICS.

SIMILAR CONICS.

157. DEF. *Conics which have the same eccentricity are said to be similar to each other.*

This definition is justified by the consideration that the character of the conic depends on its eccentricity alone, while the dimensions of all parts of the conic are entirely determined by the distance of the focus from the directrix.

Hence, according to this definition, all parabolas are similar curves.

PROP. I. *If radii be drawn from the vertices of two parabolas making equal angles with the axis, these radii are always in the same proportion.*

Let AP, ap be the radii, PN and pn the ordinates, the angles PAN, pan, being equal.

Then $$AP^2 : ap^2 :: PN^2 : pn^2 :: AS \,.\, AN : as \,.\, an.$$

But $$AP : ap :: AN : an;$$

$$\therefore AP : ap :: AS : as.$$

It can also be shewn that focal radii making equal angles with the axes are always in the same proportion.

158. PROP. II. *If two ellipses be similar their axes are in the same proportion, and any other diameters, making equal angles with the respective axes, are in the proportion of the axes.*

Let CA, CB be the semi-axes of one ellipse, ca, cb of the other, and CP, cp two radii such that the angle $PCA = pca$.

Then, since the eccentricities are the same, we have, if S, s be foci,

$$AC : SC :: ac : sc;$$

$$\therefore AC^2 : AC^2 - SC^2 :: ac^2 : ac^2 - sc^2,$$

or

$$AC^2 : BC^2 :: ac^2 : bc^2.$$

Hence it follows, if PN, pn be ordinates, that

$$PN^2 : AC^2 - CN^2 :: pn^2 : ac^2 - cn^2;$$

but, by similar triangles,

$$PN : pn :: CN : cn,$$

therefore

$$CN^2 : AC^2 - CN^2 :: cn^2 : ac^2 - cn^2;$$

and

$$CN^2 : AC^2 :: cn^2 : ac^2.$$

Hence

$$CP : cp :: CN : cn$$
$$:: AC : ac.$$

So also lines drawn similarly from the foci, or any other corresponding points of the two figures, will be in the ratio of the transverse axes.

Exactly the same demonstration is applicable to the hyperbola, but in this case, if the ratio of SC to AC in two hyperbolas be the same, it follows from Art. (102) that the angle between the asymptotes is the same in both curves.

In the case of hyperbolas we have thus a very simple test of similarity.

The Areas bounded by Conics.

159. PROP. III. *If AB, AC be two tangents to a parabola, the area between the curve and the chord BC is two-thirds of the triangle ABC.*

Draw the tangent DPE parallel to BC; then

$$AP = PN,$$

and

$$BC = 2 . DE;$$

therefore the triangle

$$BPC = 2ADE.$$

Again, draw the diameter DQM meeting BP in M.

By the same reasoning, FQG being the tangent parallel to BP, the triangle $PQB = 2FDG$.

Through F draw the diameter FRL, meeting PQ in L, and let this process be continued indefinitely.

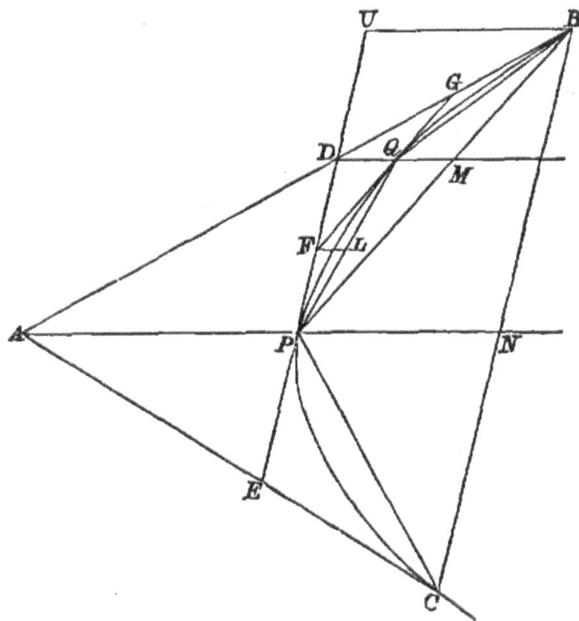

Then the sum of the triangles within the parabola is double the sum of the triangles without it.

But, since the triangle BPC is half ABC, it is greater than half the parabolic area $BQPC$;

Therefore (Euclid, Bk. XII.) the difference between the parabolic area and the sum of the triangles can be made ultimately less than any assignable quantity;

And, the same being true of the outer triangles, it follows that the area between the curve and BC is double of the area between the curve and AB, AC, and is therefore two-thirds of the triangle ABC.

COR. Since PN bisects every chord parallel to BC, it bisects the parabolic area BPC; therefore, completing the parallelogram $PNBU$, the parabolic area BPN is two-thirds of the parallelogram UN.

160. PROP. IV. *The area of an ellipse is to the area of the auxiliary circle in the ratio of the conjugate to the transverse axis.*

Draw a series of ordinates, QPN, $Q'P'N'$, ... near each other, and draw PR, QR' parallel to AC.

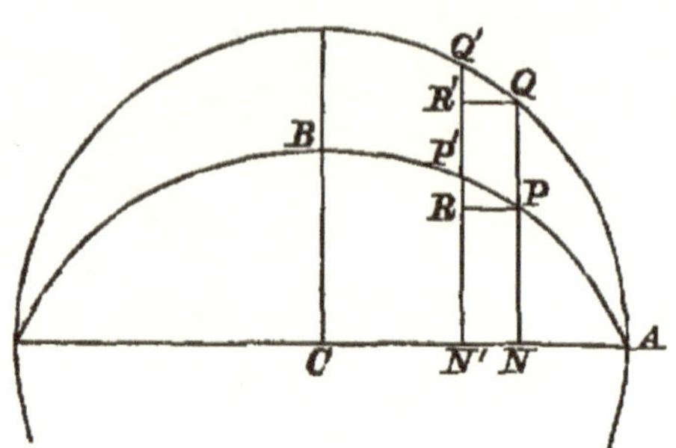

Then, since

$$PN : QN :: BC : AC,$$

the area $$PN' : QN' :: BC : AC,$$

and, this being true for all such areas, the sum of the parallelograms PN' is to the sum of the parallelograms QN' as BC to AC.

But, if the number be increased indefinitely, the sums of these parallelograms ultimately approximate to the areas of the ellipse and circle.

Hence the ellipse is to the circle in the ratio of BC to AC.

The student will find in Newton's 2nd and 3rd Lemmas (*Principia*, Section I.) a formal proof of what we have here assumed as sufficiently obvious, that the sum of the parallelograms PN is ultimately equal to the area of the ellipse.

161. Prop. V. *If P, Q be two points of an hyperbola, and if PL, QM parallel to one asymptote meet the other in L and M, the hyperbolic sector CPQ is equal to the hyperbolic trapezium $PLMQ$.*

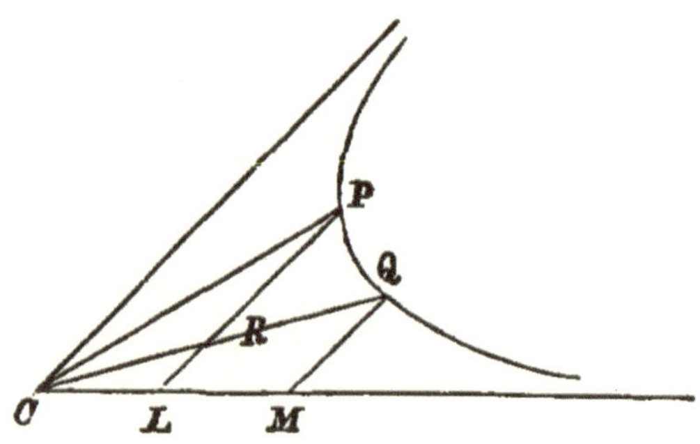

For the triangles CPL, CQM are equal, and, if PL meet CQ in R, it follows that the triangle CPR = the trapezium $LRQM$; hence, adding to each the area RPQ, the theorem is proved.

162. PROP. VI. *If points L, M, N, K be taken in an asymptote of an hyperbola, such that*

$$CL : CM :: CN : CK,$$

and if LP, MQ, NR, KS, parallel to the asymptote, meet the curve in P, Q, R, S, the hyperbolic areas CPQ, CRS will be equal.

Let QR and PS produced meet the asymptotes in F, F', G, G';

then $\qquad RF = QF'$ and $SG = PG'$ (Art. 121),

$$\therefore NF = CM \text{ and } KG = CL.$$

Hence $\qquad NF : KG :: CM : CL$

$$:: CK : CN$$

$$:: RN : SK,$$

and therefore SP is parallel to QR.

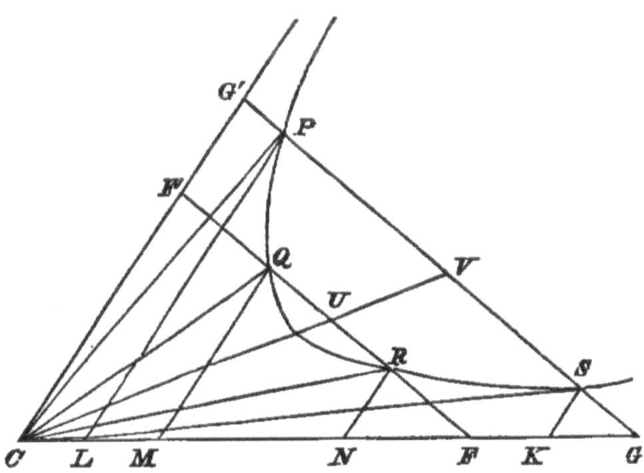

The diameter CUV conjugate to PS bisects all chords parallel to PS, and therefore bisects the area $PQRS$;

also the triangle $\qquad CPV = CSV$,

and $\qquad CQU = CUR$;

therefore, taking from CPV and CSV the equal triangles CQU, CRU, and the equal areas $PQUV$, $SRUV$, the remaining areas, which are the hyperbolic sectors CPQ, CRS, are equal.

COR. Hence if a series of points, L, M, N, ... be taken such that CL, CM, CN, CK, ... are in continued proportion, it follows that the hyperbolic sectors CPQ, CQR, CRS, &c. will be all equal.

It will be noticed in this case that the tangent at Q will be parallel to PR, the tangent at R parallel to QS, and so also for the rest.

The Curvature of Conics.

163. DEF. If a circle touch a conic at a point P, and pass through another point Q of the conic, and if the point Q move near to, and ultimately coincide with P, the circle in its ultimate condition is called the circle of curvature at P.

PROP. VII. *The chord of intersection of a conic with the circle of curvature at any point is inclined to the axis at the same angle as the tangent at the point.*

It has been shewn that, if a circle intersect a conic in four points P, Q, R, V, the chords PQ, RV are equally inclined to the axis.

Let P and Q coincide with each other; then the tangent at P and the chord RV are equally inclined to the axis.

Let the point V now approach to and coincide with P; the circle becomes the circle of curvature at P, and the chord VR becomes PR the chord of intersection.

Hence PR and the tangent at P are equally inclined to the axis.

164. PROP. VIII. *If the tangent at any point P of a parabola meet the axis in T, and if the circle of curvature at P meet the curve in Q,*

$$PQ = 4 \,.\, PT.$$

Draw the ordinate PNP'; then taking the figure of the next article, TP' is the tangent at P',

and the angle $$P'TF = PTF = PFT;$$

therefore PQ is parallel to TP', and is bisected by the diameter $P'E$.

Hence $$PQ = 2 \,.\, PE = 4P'T = 4PT.$$

165. PROP. IX. *To find the chord of curvature through the focus and the diameter of curvature at any point of a parabola.*

Let the circle meet PS produced in V, and the normal PG produced in O.

The angle $$PFS = PTS = SPT$$ $$= PQV,$$

since PT is a tangent to the circle.

Therefore QV is parallel to the axis,

and $$PV : SP :: PQ : PF.$$

Hence $$PV = 4 \,.\, SP.$$

Again, the angle $POQ = PVQ = PSN$;

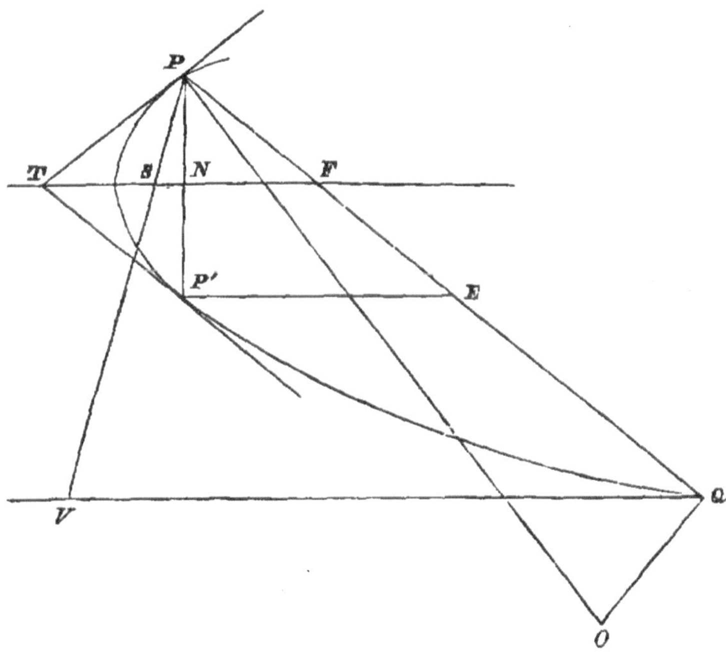

$$\therefore PO : PQ :: SP : PN,$$

or $$PO : SP :: 4PT : PN$$ $$:: 4SP : SY,$$

if SY be perpendicular to PT.

Cor. 1. Since the normal bisects the angle between SP and the diameter through P, it follows that the chord of curvature parallel to the axis is $4SP$.

Cor. 2. The diameter of curvature, PO, may also be expressed as follows:

Let GL be the perpendicular from G on SP;
then PL = the semi-latus rectum $= 2AS$.

Also PVO being a right angle,

$$PO : PG :: PV : PL$$
$$:: 4SP : PL$$
$$:: 4SP \,.\, PL : PL^2;$$

but $$4SP \,.\, PL = 8SP \,.\, AS = 8SY^2 = 2PG^2;$$
$$\therefore PO : PG :: 2PG^2 : PL^2.$$

166. Prop. X. *If the chord of intersection, PQ, of an ellipse, or hyperbola, with the circle of curvature at P, meet CD, the semi-diameter conjugate to CP, in K,*

$$PQ \,.\, PK = 2CD^2.$$

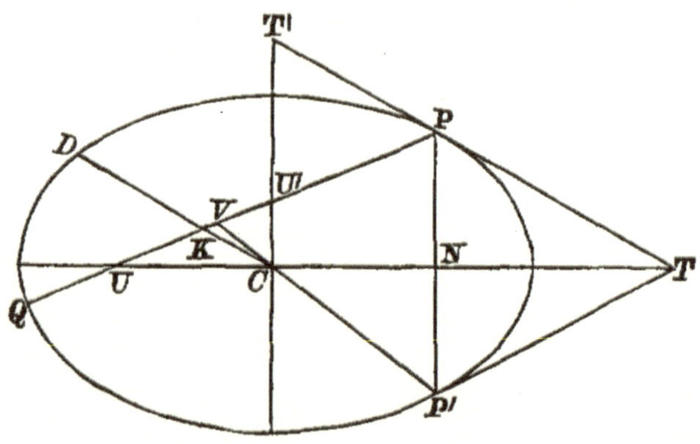

Drawing the ordinate PNP', the tangent at P' is parallel to PQ, as in the parabola, and PQ is therefore bisected in V, by the diameter CP'.

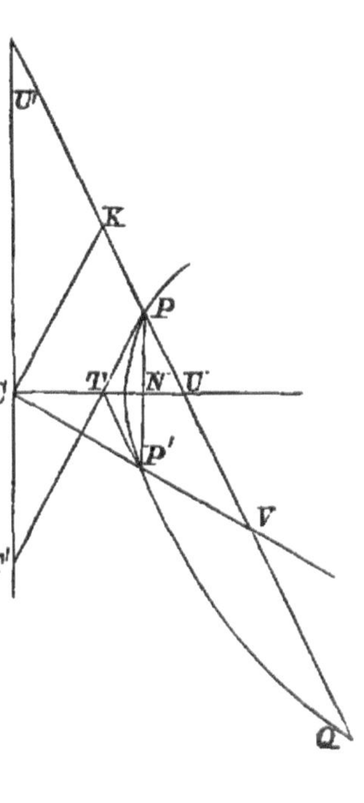

Let PQ meet the axes in U and U'; then, $U'C$ being parallel to PP',

$$PV : PU' :: VP' : CP'$$
$$:: UT : CT,$$

since PU, $P'T$ are parallel.

Also

$$UT : CT :: PU : PK;$$
$$\therefore PV : PU' :: PU : PK.$$

Hence

$$PV \,.\, PK = PU \,.\, PU'$$
$$= PT \,.\, PT' = CD^2,$$

observing that $PU = PT$, and $PU' = PT'$, by the theorem of Art. 163,

and $$\therefore PQ \,.\, PK = 2CD^2.$$

167. Prop. XI. *If the chord of curvature PQ', of an ellipse or hyperbola in any direction, meet CD in K',*

$$PQ' \,.\, PK' = 2CD^2.$$

Let PO be the diameter of curvature meeting CD in F: then PQO, $PQ'O$ are right angles, and a circle can be drawn through $Q'K'FO$;

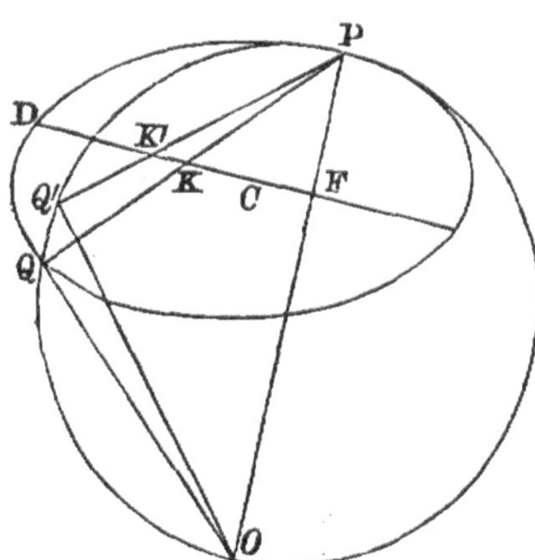

$$\therefore PQ' \,.\, PK' = PF \,.\, PO$$
$$= PK \,.\, PQ = 2 \,.\, CD^2.$$

COR. 1. Hence PO being the diameter of curvature,

$$PF \,.\, PO = 2 \,.\, CD^2.$$

COR. 2. If PQ' pass through the focus,

$$PK' = AC,$$

and
$$PQ' \,.\, AC = 2 \,.\, CD^2.$$

COR. 3. If PQ' pass through the centre,

$$PQ' \,.\, CP = 2 \,.\, CD^2.$$

168. We can also express the diameter of curvature as follows:

PG being the normal, let GL be perpendicular to SP, and let PR be the chord of curvature through S.

Then GL is parallel to OR,

and
$$PO : PG :: PR : PL$$
$$:: PR \,.\, PL : PL^2.$$

But
$$PR \,.\, AC = 2 \,.\, CD^2;$$
$$\therefore PR : AC :: 2 \,.\, CD^2 : AC^2$$
$$:: 2 \,.\, PG^2 : BC^2,$$

and
$$PR \,.\, PL : AC \,.\, PL :: 2 \,.\, PG^2 : BC^2.$$

But, PL being equal to the semi-latus rectum,

$$PL \,.\, AC = BC^2;$$
$$\therefore PR \,.\, PL = 2 \,.\, PG^2,$$

and
$$PO : PG :: 2PG^2 : PL^2.$$

Hence, in any conic, the radius of curvature at any point is to the normal at the point as the square of the normal to the square of the semi-latus rectum.

169. PROP. XII. *The chord of curvature through the focus at any point is equal to the focal chord parallel to the tangent at the point.*

Since
$$PQ' \,.\, AC = 2CD^2,$$

it follows that
$$PQ' \,.\, AA' = DD'^2.$$

But, if pp' is the focal chord parallel to the tangent at P,

$$pp' \,.\, AA' = DD'^2 \text{ (Art. 81)},$$
$$\therefore PQ' = pp'.$$

EXAMPLES.

1. The radius of curvature at the end of the latus rectum of a parabola is equal to twice the normal.

2. The circle of curvature at the end of the latus rectum intersects the parabola on the normal at that point.

3. If PV is the chord of curvature through the focus, what is the locus of the point V?

4. An ellipse and a parabola, whose axes are parallel, have the same curvature at a point P and cut one another in Q; if the tangent at P meets the axis of the parabola in T prove that $PQ = 4 \,.\, PT$.

5. In a rectangular hyperbola, the radius of curvature at P varies as CP^3.

6. If P be a point of an ellipse equidistant from the axis minor and one of the directrices, the circle of curvature at P will pass through one of the foci.

7. If the normal at a point P of a parabola meet the directrix in L, the radius of curvature at P is equal to $2 \,.\, PL$.

8. The normal at any point P of a rectangular hyperbola meets the curve again in Q; shew that PQ is equal to the diameter of curvature at P.

9. In the rectangular hyperbola, if CP be produced to Q, so that $PQ = CP$, and QO be drawn perpendicular to CQ to intersect the normal in O, O is the centre of curvature at P.

10. At any point of an ellipse the chord of curvature PV through the centre is to the focal chord pp', parallel to the tangent, as the major axis is to the diameter through the point.

11. If the common tangent of an ellipse and its circle of curvature at P be bisected by their common chord, prove that

$$CD^2 = AC \,.\, BC.$$

12. The tangent at a point P of an ellipse whose centre is C meets the axes in T and t; if CP produced meet in L the circle described about the triangle TCt, shew that PL is half the chord of curvature at P in the direction of C, and that the rectangle contained by CP, CL, is constant.

13. If P be a point on a conic, Q a point near it, and if QE, perpendicular to PQ, meet the normal at P in E, then ultimately when Q coincides with P, PE is the diameter of curvature at P.

14. If a tangent be drawn from any point of a parabola to the circle of curvature at the vertex, the length of the tangent will be equal to the abscissa of the point measured along the axis.

15. The circle of curvature at a point where the conjugate diameters are equal, meets the ellipse again at the extremity of the diameter.

16. The chord of curvature at P perpendicular to the major axis is to PM, the ordinate at P, :: $2 . CD^2 : BC^2$.

17. Prove that there is a point P on an ellipse such that if the normal at P meet the ellipse in Q, PQ is a chord of the circle of curvature at P, and find its position.

18. The chord of curvature at a point P of a rectangular hyperbola, perpendicular to an asymptote, is to CD :: $CD : 2 . PN$, where PN is the distance of P from the asymptote.

19. If G be the foot of the normal at a point P of an ellipse, and GK, perpendicular to PG, meet CP in K, then KE, parallel to the axis minor, will meet PG in the centre of curvature at P.

20. The chord of curvature through the vertex at a point of a parabola is to $4PY$ :: $PY : AP$.

21. Prove that the locus of the middle points of the common chords of a given parabola and its circles of curvature is a parabola, and that the envelope of the chords is also a parabola.

22. The circles of curvature at the extremities P, D of two conjugate diameters of an ellipse meet the ellipse again in Q, R, respectively, shew that PR is parallel to DQ.

23. The tangent at any point P in an ellipse, of which S and H are the foci, meets the axis major in T, and TQR bisects HP in Q and meets SP in R; prove that PR is one-fourth of the chord of curvature at P through S.

24. An ellipse, a parabola, and an hyperbola, have the same vertex and the same focus; shew that the curvature, at the vertex, of the parabola is greater than that of the hyperbola, and less than that of the ellipse.

25. The circle of curvature at a point of an ellipse cuts the curve in Q; the tangent at P is met by the other common tangent, which touches the curves at E and F, in T; if PQ meet TEF in O, $TEOF$ is cut harmonically.

26. If E is the centre of curvature at the point P of a parabola,

$$SE^2 + 3 \,.\, SP^2 = PE^2.$$

27. Find the locus of the foci of the parabolas which have a given circle as circle of curvature, at a given point of that circle.

28. Two parabolas, whose latera recta have a constant ratio, and whose foci are two given points A, B, have a contact of the second order at P. Shew that the locus of P is a circle.

29. If the fixed straight line PQ is the chord of an ellipse, and is also the diameter of curvature at P, prove that the locus of the centre of the ellipse is a rectangular hyperbola, the transverse axis of which is coincident in direction with PQ, and equal in length to one-half of PQ.

CHAPTER VIII.

Orthogonal Projections.

170. Def. The projection of a point on a plane is the foot of the perpendicular let fall from the point on the plane.

If from all points of a given curve perpendiculars be let fall on a plane, the curve formed by the feet of the perpendiculars is the projection of the given curve.

The projection of a straight line is also a straight line, for it is the line of intersection with the given plane of a plane through the line perpendicular to the given plane.

Parallel straight lines project into parallel lines, for the projections are the lines of intersection of parallel planes with the given plane.

171. Prop. I. *Parallel straight lines, of finite lengths, are projected in the same ratio.*

That is, if ab, pq be the projections of the parallel lines AB, PQ,

$$ab : AB :: pq : PQ.$$

For, drawing AC parallel to ab and meeting Bb in C, and PR parallel to pq and meeting Qq in R, ABC and PQR are similar triangles; therefore

$$AC : AB :: PR : PQ,$$

and

$$AC = ab,\ PR = pq.$$

172. Prop. II. *The projection of the tangent to a curve at any point is the tangent to the projection of the curve at the projection of the point.*

For if p, q be the projections of the two points P, Q of a curve, the line pq is the projection of the line PQ, and when the line PQ turns round P until Q coincides with P, pq turns round p until q coincides with p, and the ultimate position of pq is the tangent at p.

173. PROP. III. *The projection of a circle is an ellipse.* Let aba' be the projection of a circle ABA'.

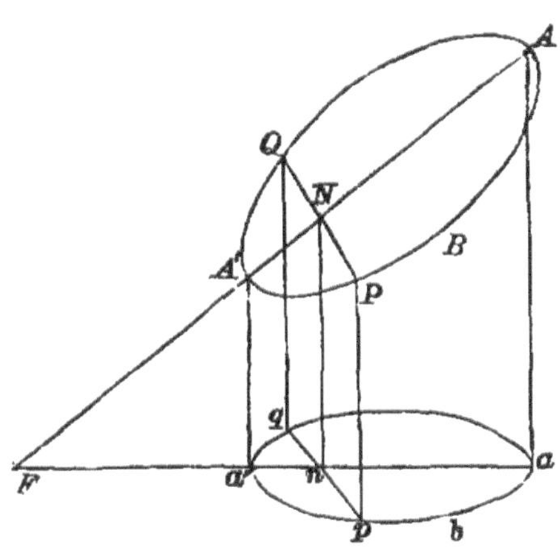

Take a chord PQ parallel to the plane of projection, then its projection $pq = PQ$.

Let the diameter ANA' perpendicular to PQ meet in F the plane of projection, and let $aa'F$ be the projection of $AA'F$.

Then aa' bisects pq at right angles in the point n, and

$$an : AN :: aF : AF,$$

$$a'n : A'N :: aF : AF;$$

$$\therefore AN \,.\, NA' : an \,.\, na' :: AF^2 : aF^2;$$

but

$$AN \,.\, NA' = PN^2 = pn^2,$$

$$\therefore pn2 : an \,.\, na' :: AF^2 : aF^2,$$

and the curve apa' is an ellipse, having its axes in the ratio of

$$aF : AF, \quad \text{or of} \quad aa' : AA'.$$

Moreover, since we can place the circle so as to make the ratio of aa' to AA' whatever we please, an ellipse of any eccentricity can be obtained.

In this demonstration we have assumed only the property of the principal diameters of an ellipse. Properties of other diameters can be obtained by help of the preceding theorems, as in the following instances.

174. PROP. IV. *The locus of the middle points of parallel chords of an ellipse is a straight line.*

For, projecting a circle, the parallel chords of the ellipse are the projections of parallel chords of the circle, and as the middle points of these latter lie in a diameter of the circle, the middle points of the chords of the ellipse lie in the projection of the diameter, which is a straight line, and is a diameter of the ellipse.

Moreover, the diameter of the circle is perpendicular to the chords it bisects; hence

Perpendicular diameters of a circle project into conjugate diameters of an ellipse.

175. PROP. V. *If two intersecting chords of an ellipse be parallel to fixed lines, the ratio of the rectangles contained by their segments is constant.*

Let OPQ, ORS be two chords of a circle, parallel to fixed lines, and opq, ors their projections.

Then $OP \,.\, OQ$ is to $op \,.\, oq$ in a constant ratio, and $OR \,.\, OS$ is to $or \,.\, os$ in a constant ratio; but

$$OP \,.\, OQ = OR \,.\, OS.$$

Therefore $op \,.\, oq$ is to $or \,.\, os$ in a constant ratio; and opq, ors are parallel to fixed lines.

176. PROP. VI. *If qvq' be a double ordinate of a diameter cp, and if the tangent at q meet cp produced in t,*

$$cv \,.\, ct = cp^2.$$

The lines qvq' and cp are the projections of a chord QVQ' of a circle which is bisected by a diameter CP, and t is the projection of T the point in which the tangent at Q meets CP produced.

But, in the circle,

$$CV \,.\, CT = CP^2,$$

or

$$CV : CP :: CP : CT;$$

and, these lines being projected in the same ratio, it follows that

$$cv : cp :: cp : ct,$$

or

$$cv \,.\, ct = cp^2.$$

Hence it follows that tangents to an ellipse at the ends of any chord meet in the diameter conjugate to the chord.

The preceding articles will shew the utility of the method in dealing with many of the properties of an ellipse.

The student will find it useful to prove, by orthogonal projections, the theorems of Arts. 58, 69, 74, 75, 78, 79, 80, 82, 83, 89, 90, and 92.

177. PROP. VII. *An ellipse can be projected into a circle.*

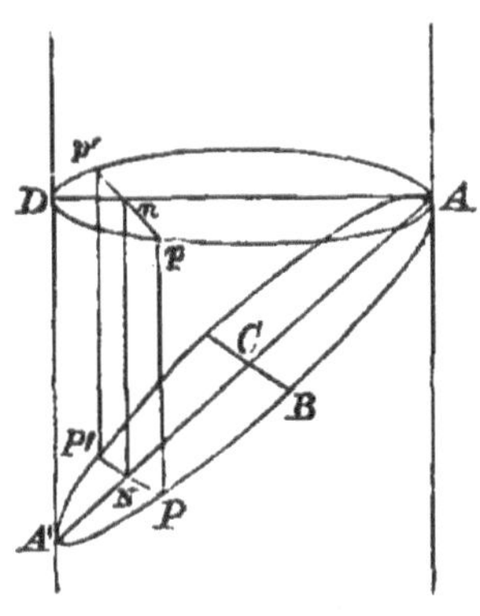

This is really the converse of Art. 173, but we give a construction for the purpose.

Draw a plane through AA', the transverse axis, perpendicular to the plane of the ellipse, and in this plane describe a circle on AA' as diameter. Also take the chord AD, equal to the conjugate axis, and join $A'D$, which is perpendicular to AD.

Through AD draw a plane perpendicular to $A'D$, and project a principal chord PNP' on this plane.

Then $$PN^2 : AN \,.\, NA' :: BC^2 : AC^2.$$

But $$PN = pn,$$

$$An : AN :: AD : AA'$$

$$:: BC : AC,$$

and $$Dn : A'N :: BC : AC.$$

Hence $$An \,.\, nD : AN.NA' :: BC^2 : AC^2,$$

and therefore $$pn^2 = An \,.\, nD,$$

and the projection ApD is a circle.

This theorem, in the same manner as that of Art. 173, may be employed in deducing properties of oblique diameters and oblique chords of an ellipse.

178. *If any figures in one plane be projected on another plane, the areas of the projections will all be in the same ratio to the areas of the figures themselves.*

Let BAD be the plane of the figures, and let them be projected on the plane CAD, C being the projection of the point B, and BAD being a right angle.

Taking a rectangle $EFGH$, the sides of which are parallel and perpendicular to AD, the projection is $efgh$, and it is clear that the ratio of the areas of these rectangles is that of AC to AB.

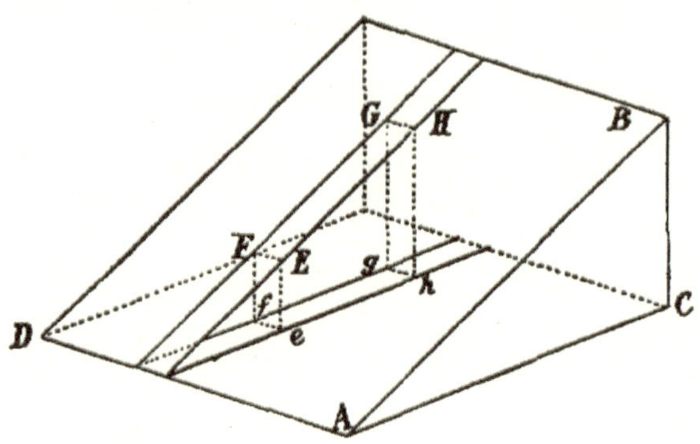

Now the area of any curvilinear figure in the plane BAD is the sum of the areas of parallelograms such as $EFGH$, which are inscribed in the figure, if we take the widths, such as EF, infinitesimally small.

It follows that the area of the projection of the figure is to the area of the figure itself in the ratio of AC to AB.

As an illustration, let a square be drawn circumscribing a circle, and project the figure on any plane. The square projects into tangents parallel to conjugate diameters of the ellipse which is the projection of the circle.

The area of the parallelogram thus formed is the same whatever be the position of the square, and we thus obtain the theorem of Art. 87.

179. It follows that maxima and minima areas project into maxima and minima areas. For example, the greatest triangle which can be inscribed in a circle is an equilateral triangle.

Projecting this figure we find that the triangle of maximum area inscribed in an ellipse is such that the tangent at each angular point is parallel to the opposite side, and that the centre of the ellipse is the point of intersection of the lines joining the vertices of the triangle with the middle points of the opposite sides.

180. PROP. VII. *The projection of a parabola is a parabola.*

For if PNP' be a principal chord, bisected by the axis AN, the projection pnp' will be bisected by the projection an.

Moreover $pn : PN$ will be a constant ratio, as also will be $an : AN$.

And $$PN^2 = 4AS \,.\, AN.$$

Hence pn^2 will be to $4AS \,.\, an$ in a constant ratio, and the projection is a parabola, the tangent at a being parallel to pn.

181. PROP. VIII. *An hyperbola can be always projected into a rectangular hyperbola.*

For the asymptotes can be projected into two straight lines cl, cl' at right angles, and if PM, PN be parallels to the asymptotes from a point P of the curve, $PM \,.\, PN$ is constant.

But $pm : PM$ and $pn : PN$ are constant ratios;

$$\therefore pm \,.\, pn \text{ is constant.}$$

And since pm and pn are perpendicular respectively to cl and cl', it follows that the projection is a rectangular hyperbola.

The same proof evidently shews that any projection of an hyperbola is also an hyperbola.

EXAMPLES.

1. A parallelogram is inscribed in a given ellipse; shew that its sides are parallel to conjugate diameters, and find its greatest area.

2. TP, TQ are tangents to an ellipse, and CP', CQ' are parallel semi-diameters; PQ is parallel to $P'Q'$.

3. If a straight line meet two concentric similar and similarly situated ellipses, the portions intercepted between the curves are equal.

4. Find the locus of the point of intersection of the tangents at the extremities of pairs of conjugate diameters of an ellipse.

5. Find the locus of the middle points of the lines joining the extremities of conjugate diameters.

6. If a tangent be drawn at the extremity of the major axis meeting two equal conjugate diameters CP, CD produced in T and t; then $PD^2 = 2AT^2$.

7. If a chord AQ drawn from the vertex be produced to meet the minor axis in O, and CP be a semi-diameter parallel to it, then $AQ \,.\, AO = 2CP^2$.

8. OQ, OQ' are tangents to an ellipse from an external point O, and OR is a diagonal of the parallelogram of which OQ, OQ' are adjacent sides; prove that if R be on the ellipse, O will lie on a similar and similarly situated concentric ellipse.

9. AB is a given chord of an ellipse, and C any point in the ellipse; shew that the locus of the point of intersection of lines drawn from A, B, C to the middle points of the opposite sides of the triangle ABC is a similar ellipse.

10. CP, CD are conjugate semi-diameters of an ellipse; if an ellipse, similar and similarly situated to the given ellipse, be described on PD as diameter, it will pass through the centre of the given ellipse.

11. Parallelograms are inscribed in an ellipse and one pair of opposite sides constantly touch a similar, similarly situated and concentric ellipse; shew that the remaining pair of sides are tangents to a third ellipse and the square on a principal semi-axis of the original ellipse is equal to the sum of the squares on the corresponding semi-axes of the other two ellipses.

12. Find the locus of the middle point of a chord of an ellipse which cuts off a constant area from the curve.

13. Find the locus of the middle point of a chord of a parabola which cuts off a constant area from the curve.

14. A parallelogram circumscribes an ellipse, touching the curve at the extremities of conjugate diameters, and another parallelogram is formed by joining the points where its diagonals meet the ellipse: prove that the area of the inner parallelogram is half that of the outer one.

If four similar and similarly situated ellipses be inscribed in the spaces between the outer parallelogram and the curve, prove that their centres lie in a similar and similarly situated ellipse.

15. About a given triangle PQR is circumscribed an ellipse, having for centre the point of intersection (C) of the lines from P, Q, R bisecting the opposite sides, and PC, QC, RC are produced to meet the curve in P', Q', R'; shew that, if tangents be drawn at these points, the triangle so formed will be similar to PQR, and four times as great.

16. The locus of the middle points of all chords of an ellipse which pass through a fixed point in an ellipse similar and similarly situated to the given ellipse, and with its centre in the middle point of the line joining the given point and the centre of the given ellipse.

17. PT, pt are tangents at the extremities of any diameter Pp of an ellipse; any other diameter meets PT in T and its conjugate meets pt in t; also any tangent meets PT in T' and pt in t'; shew that $PT : PT' :: pt' : pt$.

18. From the ends P, D of conjugate diameters of an ellipse lines are drawn parallel to any tangent line; from the centre C any line is drawn cutting these lines and the tangent in p, d, t, respectively; prove that $Cp^2 + Cd^2 = Ct^2$.

19. If CP, CD be conjugate diameters of an ellipse, and if BP, BD be joined, and also AD, $A'P$, these latter intersecting in O, the figure $BDOP$ will be a parallelogram.

20. T is a point on the tangent at a point P of an ellipse, so that a perpendicular from T on the focal distance SP is of constant length; shew that the locus of T is a similar, similarly situated and concentric ellipse.

21. Q is a point in one asymptote, and q in the other. If Qq move parallel to itself, find the locus of intersection of tangents to the hyperbola from Q and q.

22. Tangents are drawn to an ellipse from an external point T. The chord of contact and the major axis, or these produced, intersect in K, and TN is drawn perpendicular to the major axis. Prove that

$$CN\,.\,CK = CA^2.$$

23. Q is a variable point on the tangent at a fixed point P of an ellipse and R is taken so that $PQ = QR$. If the other tangent from Q meet the ellipse in K, prove that RK passes through a fixed point.

24. If through any point on an ellipse there be drawn lines conjugate to the sides of an inscribed triangle they will meet the sides in three points in a straight line.

25. PCP' is a diameter of an ellipse, and a chord PQ meets the tangent at P' in R. Prove that PQ, PR have the parallel diameter for a mean proportional.

26. If AOA', BOB' are conjugate diameters of an ellipse, and if AP and BQ are parallel chords, $A'Q$ and $B'P$ are parallel to conjugate diameters.

27. If the tangents at the ends of a chord of an hyperbola meet in T, and TM, TM' be drawn parallel to the asymptotes to meet them in M, M', then MM' is parallel to the chord.

28. If a windmill in a level field is working uniformly on a sunny day, the speed of the end of the shadow of one sail varies as the length of the shadow of the next sail.

29. Spheres are drawn passing through a fixed point and touching two fixed planes. Prove that the points of contact lie on two circles, and that the locus of the centre of the sphere is an ellipse.

If the angle between the planes is the angle of an equilateral triangle, prove that the distance between the foci of the ellipse is half the major axis.

CHAPTER IX.

Of Conics in General.

The Construction of a Conic.

182. The method of construction, given in Chapter I., can be extended in the following manner.

Let fSn be any straight line drawn through the focus S, and draw Ax from the vertex parallel to fS, and meeting the directrix in x.

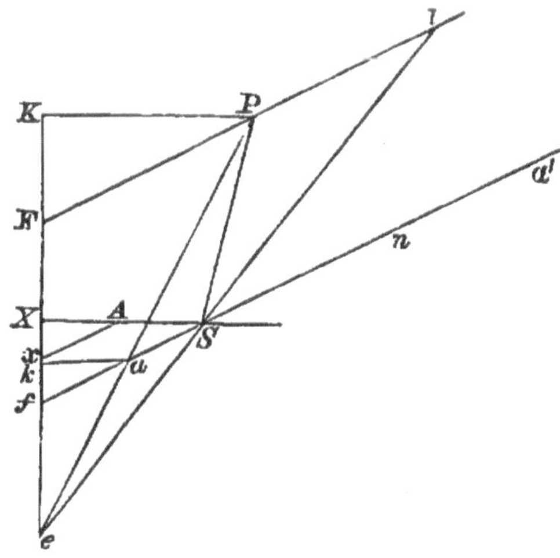

Divide the line fSn in a and a' so that

$$Sa : af :: Sa' : a'f :: SA : Ax;$$

then a and a' are points on the curve, for, if ak be the perpendicular on the directrix,

$$ak : af :: AX : Ax,$$

and therefore $$Sa : ak :: SA : AX.$$

Take any point e in the directrix, draw the lines eSl, ea through S and a, and draw SP making the angle PSl equal to lSn.

Through P draw FPl parallel to fS, and meeting eS produced in l,

then $$Pl = SP,$$

and $$Pl : PF :: Sa : af;$$

$$\therefore SP : PF :: Sa : af,$$

and $$SP : PK :: Sa : ak;$$

therefore P is a point in the curve.

183. The construction for the point a gives a simple proof that the tangent at the vertex is perpendicular to the axis. For when the angle ASa is diminished, Sa approaches to equality with SA, and therefore the angle aAS is ultimately a right angle.

184. Prop. I. *To find the points in which a given straight line is intersected by a conic of which the focus, the directrix, and the eccentricity are given.*

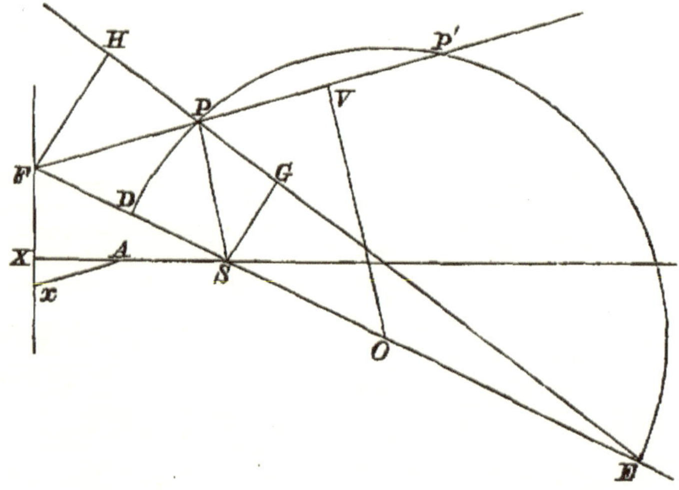